Advancing Maths for AQA
MECHANICS 2

Ted Graham and Aidan Burrows

Series editors
Ted Graham   Sam Boardman
David Pearson   Roger Williamson

1	Moments and equilibrium	1
2	Centres of mass	16
3	Energy	32
4	Kinematics and variable acceleration	68
5	Circular motion	91
6	Circular motion with variable speed	112
7	Application of differential equations in mechanics	126
Exam style practice papers		134
Answers		140
Index		153

heinemann.co.uk
✓ Free online support
✓ Useful weblinks
✓ 24 hour online ordering

01865 888058

Heinemann
Inspiring generations

Heinemann is an imprint of Pearson Education Limited, a company incorporated in England and Wales, having its registered office at Edinburgh Gate, Harlow, Essex, CM20 2JE. Registered company number: 872828

Heinemann is a registered trademark of
Pearson Education Limited

© Ted Graham, Aidan Burrows and Joan Corbett 2000, 2004
Complete work © Heinemann Educational Publishers 2004

First published 2004

08
10 9 8 7 6 5 4

British Library Cataloguing in Publication Data is available
from the British Library on request.

ISBN: 978 0 435513 37 5

Copyright notice
All rights reserved. No part of this publication may be reproduced in any form or by any means (including photocopying or storing it in any medium by electronic means and whether or not transiently or incidentally to some other use of this publication) without the written permission of the copyright owner, except in accordance with the provisions of the Copyright, Designs and Patents Act 1988 or under the terms of a licence issued by the Copyright Licensing Agency, 90 Tottenham Court Road, London W1T 4LP. Applications for the copyright owner's written permission should be addressed to the publisher.

Edited by Alex Sharpe, Standard Eight Limited
Typeset and illustrated by Tech-Set Limited, Gateshead, Tyne & Wear
Original illustrations © Harcourt Education Limited, 2004
Cover design by Miller, Craig and Cocking Ltd
Printed in China (SWTC/04)

Acknowledgements
The publishers and authors acknowledge the work of the writers Ray Atkin, John Berry, Derek Collins, Tim Cross, Ted Graham, Phil Rawlins, Tom Roper, Rob Summerson, Nigel Price, Frank Chorlton and Andy Martin of the *AEB Mathematics for AQA A-Level* Series, from which some exercises and examples have been taken.

The publishers' and authors' thanks are due to AQA for permission to reproduce questions from past examination papers.

The answers have been provided by the authors and are not the responsibility of the examining board.

Every effort has been made to contact copyright holders of material reproduced in this book. Any omissions will be rectified in subsequent printings if notice is given to the publishers.

About this book

This book is one in a series of textbooks designed to provide you with exceptional preparation for AQA's 2004 Mathematics Specification. The series authors are all senior members of the examining team and have prepared the textbooks specifically to support you in studying this course.

Finding your way around

The following are there to help you find your way around when you are studying and revising:
- **edge marks** (shown on the front page) – these help you to get to the right chapter quickly;
- **contents list** – this identifies the individual sections dealing with key syllabus concepts so that you can go straight to the areas that you are looking for;
- **index** – a number in bold type indicates where to find the main entry for that topic.

Key points

Key points are not only summarised at the end of each chapter but are also boxed and highlighted within the text like this:

> The moment of a force about a point is defined as the magnitude of the force multiplied by the perpendicular distance from the point to the force or the line of action of the force.

Exercises and exam questions

Worked examples and carefully graded questions familiarise you with the syllabus and bring you up to exam standard. Each book contains:
- Worked examples and Worked exam questions to show you how to tackle typical questions;
- Graded exercises, gradually increasing in difficulty up to exam-level questions, which are marked by an [A];
- Test-yourself sections for each chapter so that you can check your understanding of the key aspects of that chapter and identify any sections that you should review;
- Answers to the questions are included at the end of the book.

Contents

1 Moments and equilibrium

Learning objectives   1
1.1 Introduction   1
1.2 The moment of a force   1
1.3 Moments and equilibrium: part 1   4
1.4 Moments and equilibrium: part 2   9
Key point summary   15
Test yourself   15

2 Centres of mass

Learning objectives   16
2.1 Centre of mass of a system of particles   16
    Working in two dimensions   18
2.2 Centre of mass of a composite body   22
    Centre of mass of a lamina   23
    The position of a suspended lamina or body   24
Key point summary   30
Test yourself   31

3 Energy

Learning objectives   32
3.1 Kinetic energy   32
3.2 Work and energy   33
3.3 Forces at angles   38
    Gravitational potential energy   40
3.4 Hooke's law   47
3.5 Energy and variable forces   49
    Elastic potential energy   49
3.6 Power   57
    Work done by force   57
Key point summary   65
Test yourself   66

4 Kinematics and variable acceleration

Learning objectives   68
4.1 Introduction   68
4.2 Displacement to velocity and acceleration   69
4.3 Acceleration to velocity and displacement   75
4.4 Motion in two or three dimensions   81
    Obtaining a position vector   83
Key point summary   89
Test yourself   90

5 Circular motion

Learning objectives	91
5.1 Introduction	91
5.2 Angular speed	91
Velocity and circular motion	92
Acceleration and circular motion	92
5.3 Forces involved in horizontal circular motion	95
5.4 Further circular motion	101
Conical pendulum	101
Key point summary	111
Test yourself	111

6 Circular motion with variable speed

Learning objectives	112
6.1 Introduction	112
6.2 Circular motion with variable speed	112
6.3 Motion in a vertical circle	115
Key point summary	125
Test yourself	125

7 Application of differential equations in mechanics

Learning objectives	126
Key point summary	133
Test yourself	133

Exam style practice papers

	134
MM2A	134
MM2B	137

Answers
140

Index
153

CHAPTER 1
Moments and equilibrium

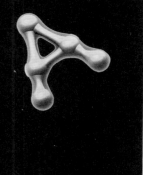

Learning objectives

After studying this chapter you should be able to:
- find the moment of a force
- know that the resultant force and moment must both be zero for a rigid body to be in equilibrium

1.1 Introduction

In this chapter we move on from modelling objects as particles and introduce the idea of a rigid body. This is a body that does not change shape, but that does have size. The major difference between a body modelled as a particle and a rigid body is that the place where the force acts on a rigid body becomes important.

Consider two forces of equal magnitude, but opposite directions that act on a book resting on a table. The diagram shows the particle model and a possible rigid body model.

The two situations would produce different results if the forces were applied as shown. The particle would remain at rest, but the rigid body would rotate. It is the way in which the forces are applied that causes the body to rotate. In the first section of this chapter we will introduce the idea of the moment of a force, which gives a measure of the turning effect of a force.

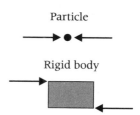

Particle

Rigid body

1.2 The moment of a force

The turning effect of a force depends on the size and direction of the force and the point where the force acts. Consider a door that is hinged at O, as shown in the diagram.

It is easier to open the door with a force applied at B than at A. The greater the distance of the force from the hinge the greater the turning effect.

2 Moments and equilibrium

> The moment of a force about a point is defined as the magnitude of the force multiplied by the perpendicular distance from the point to the force or the line of action of the force.

In this diagram the distance specified is perpendicular to the force and so the perpendicular distance is simply d.

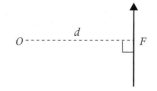

Moment of the force about $O = Fd$

If the distance specified is not perpendicular to the force, then the perpendicular distance must be calculated.

In this case the specified distance, d, is not the perpendicular distance and so this must be calculated. Here the perpendicular distance is $d \sin \theta$, so the moment about O is

$Fd \sin \theta$

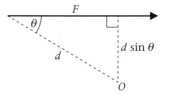

Sometimes a force has to be extended, to show the line of action of the force, so that the perpendicular distance can be found.

> A moment can have a clockwise or anticlockwise turning effect. An anticlockwise moment is defined as a positive moment and a clockwise moment is defined as a negative moment.

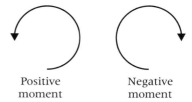

Positive moment   Negative moment

When calculating a moment, a force is multiplied by a distance, so the SI units of a moment are Newton metres or N m.

Worked example 1.1

Find the moment of each of these forces about the point O.

(a)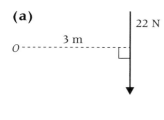

(b)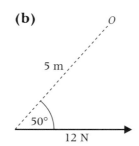

(c)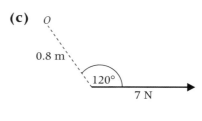

Solution

(a) Here the perpendicular distance is 3 m.

Moment about $O = -22 \times 3 = -66$ N m

The force produces a clockwise moment and so is negative.

(b) Here the perpendicular distance must be drawn in, as shown in the diagram. In this case it is 5 sin 50°.

  Moment about $O = 12 \times 5 \sin 50°$

  $ = 46.0 \text{ N m}$

The moment is anticlockwise and so is positive.

(c) To find the perpendicular distance in this case the line of action of the force must be drawn, by extending the arrow used to represent the force. The perpendicular distance can then be calculated as 0.8 sin 60°.

  Moment about $O = 7 \times 0.8 \sin 60°$

  $ = 4.85 \text{ N m}$

The moment is anticlockwise and so is positive.

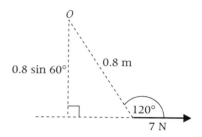

EXERCISE 1A

1 Find the moment about the point O of each of the forces below.

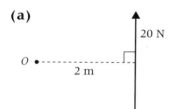
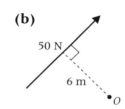
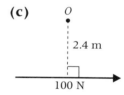

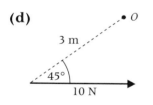
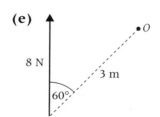
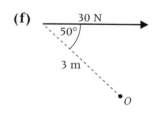

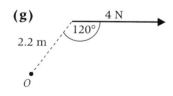
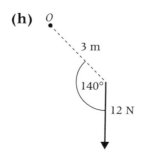
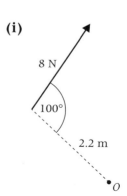

4 *Moments and equilibrium*

2 If the moment about the point *O* of each force below is −40 N m, find the distance *d* in each case.

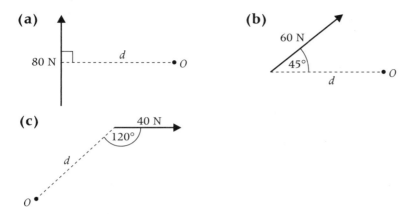

1.3 Moments and equilibrium: part 1

> For a particle to be in equilibrium, the resultant force on the particle must be zero. For a rigid body to be in equilibrium, the resultant force must be zero and the total moment of all the forces acting must also be zero.

This is illustrated in the diagram below.

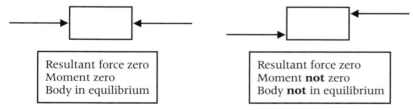

The following examples show how this principle can be applied to physical situations.

In all the examples below we will consider uniform bodies. These are bodies that are made from the same material throughout. These bodies will also be symmetric shapes so that we can assume that the force of gravity or weight acts at the centre of the object.

Worked example 1.2

A uniform beam of mass 20 kg and length 3 m rests on two supports as shown below.

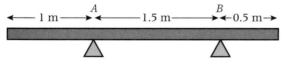

(a) Find the reaction force exerted by each support.

(b) Find the greatest mass that can be placed at the left-hand end of the beam, if it is to remain in equilibrium.

Solution

(a) The diagram shows the forces acting on the beam.

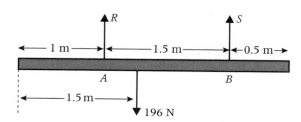

As the upward forces must balance the downward force:

$$R + S = 196$$

Taking moments about the point A gives:

$$0.5 \times 196 = 1.5S$$

$$S = \frac{98}{1.5}$$

$$= 65\frac{1}{3}\,\text{N}$$

Then using $R + S = 196$ gives:

$$R + 65\frac{1}{3} = 196$$

$$R = 130\frac{2}{3}\,\text{N}$$

(b) If a mass is placed at the left-hand end of the beam, then an extra force is added to the diagram as shown.

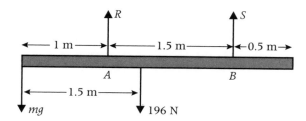

When the beam is just in equilibrium with the maximum possible mass, the reaction force, S, will be zero. The beam is effectively balanced on the support at A.

Taking moments about A and with $S = 0$ gives:

$$1 \times 9.8m = 0.5 \times 196$$

$$m = 10\,\text{kg}$$

Worked example 1.3

The diagram shows a uniform beam, AB, of length 2 m and mass 20 kg. It is supported at each end and a load of mass 100 kg is suspended from the beam in the position shown.

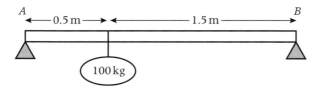

Find the magnitude of the reaction forces on each end of the beam.

Solution

The diagram shows the forces acting on the beam.

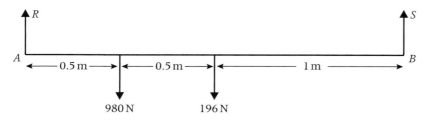

Taking moments about A gives:

$$(0.5 \times 980) + (1 \times 196) = 2 \times S$$

$$S = \frac{(0.5 \times 980) + (1 \times 196)}{2}$$

$$= 343 \text{ N}$$

As the upward forces balance the downward forces:

$$R + S = 980 + 196$$

$$R + 343 = 1176$$

$$R = 833 \text{ N}$$

EXERCISE 1B

1 A see-saw of length 4 m, pivoted at its centre, rests in a horizontal position. John, who has mass 30 kg, sits at one end. Where should his friend James, who has mass 40 kg, sit if the see-saw is to remain in a horizontal position?

2 A uniform beam of mass 30 kg and length 8 m, rests on supports that are 2 m from each end. A load of mass 50 kg is suspended from a point 3 m from one end. Find the magnitude of the reaction forces exerted by the two supports.

3 The diagram shows a uniform metal beam of mass 40 kg that rests on two supports.

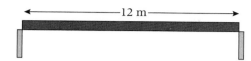

(a) Find the magnitude of the two reaction forces acting on the beam.

(b) A load of mass 30 kg is suspended from the beam, at a point 4 m from one end. Find the reaction forces at each end of the beam.

4 A beam of length 3 m and mass 15 kg is supported as shown in the diagram.

(a) Find the magnitude of each of the reaction forces acting on the beam.

(b) A mass is placed at the unsupported end of the beam. What is the greatest mass that can be placed in this position?

5 The diagram shows a uniform beam of mass 20 kg that rests on two concrete blocks.

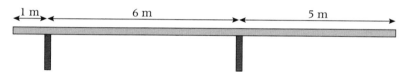

(a) Find the magnitude of each of the reaction forces acting on the beam.

(b) A box is to be placed on the beam. What is the greatest mass of a box that could be placed at the right-hand end of the beam?

6 A uniform plank of length 1.6 m and mass 4 kg rests on two supports which are 0.3 m from each end of the plank. A mass is attached to one end of the plank. If the normal reaction force on the support nearer to this load is twice the normal reaction force on the other support, determine the mass attached.

7 A uniform plank of length 2 m and mass 6 kg is suspended in a horizontal position by two vertical strings, one at each end. A 2 kg mass is placed on the plank at a variable point P. If either string snaps when the tension in it exceeds 42 N, find the section of the plank in which P can be.

8 A metre rule is pivoted 20 cm from one end A and is balanced in a horizontal position by hanging a mass of 180 grams at A. What is the mass of the rule? What additional mass should be hung from A if the pivot is moved 10 cm nearer to A?

9 A metre rule of mass 100 grams is placed on the edge of a table as shown with a 200 grams mass at A. A mass M grams is attached at C.

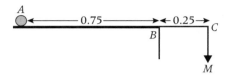

(a) When the rule is just on the point of overturning, where does the normal contact force act?

(b) Determine the maximum value of M for which the rule will not overturn.

10 A simple bridge consists of a uniform plank that is supported in a horizontal position by two vertical ropes. The plank is modelled as a rod of length 4 m and mass 60 kg. The ropes are attached to the plank at A and B, as shown in the diagram.

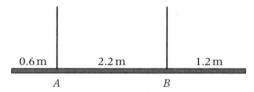

(a) Draw a diagram to show the three forces acting on the plank.

(b) Show that the tension in the rope attached to the plank at A is 214 N, correct to three significant figures.

(c) Find the tension in the rope attached to the plank at B.
[A]

11

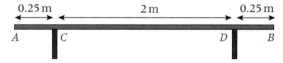

A uniform plank, AB, rests horizontally on two fixed vertical supports at C and D. The plank has mass 10 kg and length 2.5 m. The supports at C and D are 0.25 m from A and B respectively, as shown in the diagram.

Steve has mass 60 kg and stands on the plank at a point 0.75 m from A.

(a) Draw a diagram showing all the forces acting on the plank.

(b) Find the reaction of the supports on the plank. [A]

12 A uniform metal bar, of mass 30 kg and length 3 m, rests in a horizontal position on two supports at A and B, as shown in the diagram below.

Find the magnitude of each of the reactions forces acting on the bar at the supports at A and B. [A]

13 A uniform metal beam has length 5 m and mass 250 kg. It rests horizontally on two supports, A and B, which are 3 m apart. Support A is at one end of the beam, as shown in the diagram.

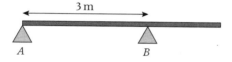

(a) Find the magnitudes of the forces exerted on the beam by the supports.

(b) A man, of mass 80 kg, walks along the beam from A towards the other end of the beam. Find the distance he can walk past B, before the beam starts to tip. [A]

1.4 Moments and equilibrium: part 2

In this section we consider cases where the forces acting are not parallel.

Worked example 1.4

The diagram shows a rod, of mass 30 kg and length 3 m, that is smoothly hinged at A. The rod is held in a horizontal position by a rope. The rope is attached to the rod at a point B, that is 2 m from A. The angle between the rope and the rod is 60°. A load, of mass 100 kg, is suspended from the end of the rod at C.

Find the tension in the rope.

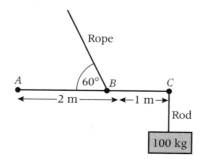

Solution

The diagram shows the forces acting. We will find the tension by taking moments about the point A.

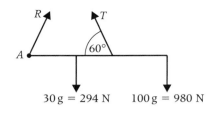

The perpendicular from A to the tension force is 2 sin 60°.

Taking moments about A gives

$$T \times 2 \sin 60° = 100 \times 9.8 \times 3 + 30 \times 9.8 \times 1.5$$

$$T = \frac{3381}{2 \sin 60°}$$

$$= 1952 \text{ N}$$

Note that because we take moments about A, we do not need to worry about the size of the reaction force, R, that acts at A.

Worked example 1.5

A ladder, of length 5 m and mass 20 kg, leans against a smooth wall so that it is at an angle of 60° to the horizontal. The ladder remains at rest, with its base on rough, horizontal ground.

(a) Find the magnitude of the normal reaction and friction forces acting on the base of the ladder.

(b) Find an inequality that the coefficient of friction must satisfy.

Solution

(a) The diagram shows the forces acting on the ladder.

As the vertical forces acting on the ladder are in equilibrium, we have:

$$R = 20g$$
$$= 196 \text{ N}$$

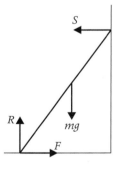

As the horizontal forces are also in equilibrium, we have:

$$F = S$$

Next we will take moments about the base of the ladder. For the weight, mg, the perpendicular distance from the base to the force is 2.5 cos 60° and the perpendicular distance from the base to the force S is 5 cos 30°. Hence taking moments about the base gives:

$$S \times 5 \cos 30° - 196 \times 2.5 \cos 60° = 0$$

$$S = \frac{196 \times 2.5 \cos 60°}{5 \cos 30°}$$

$$= 56.58 \text{ N}$$

But as $F = S$, we have $F = 56.58$ N.

(b) As the ladder is at rest the friction inequality $F \leq \mu R$ must be satisfied. Substituting for R and F gives:

$$F \leq \mu R$$
$$56.58 \leq \mu \times 196$$
$$\mu \geq \frac{56.58}{196} = 0.289 \text{ (to three significant figures)}$$

Worked example 1.6

A boom on a yacht is held in a horizontal position by a rope attached to the top of the mast. The boom is freely pivoted where it is attached to the mast. The length of the boom is 4 m and its mass is 15 kg. The rope makes an angle of 70° with the boom.

(a) Find the tension in the rope.

(b) Find the magnitude of the force that the mast exerts on the boom.

Solution

(a) The diagram shows the forces acting on the boom.

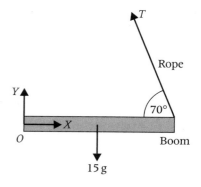

Taking moments about the point, O, where the boom is attached to the mast gives:

$$T \times 4 \sin 70° = 15 \times 9.8 \times 2$$
$$T = \frac{294}{4 \sin 70°}$$
$$= 78.2 \text{ N}$$

(b) To find the force exerted by the mast, assume that it has two components, X horizontal and Y vertical.

Then for horizontal equilibrium, we require:

$$X = T \cos 70°$$
$$= \frac{294 \cos 70°}{4 \sin 70°}$$
$$= 26.8 \text{ N}$$

For vertical equilibrium

$$Y + T \sin 70° = 15 \times 9.8$$
$$Y = 147 - \frac{294}{4}$$
$$= 73.5 \text{ N}$$

The magnitude of the force acting is given by $\sqrt{X^2 + Y^2}$.

Substituting the values obtained above gives 78.2 N.

EXERCISE 1C

1 The diagram shows the forces that act on a rectangular sheet of metal. All the forces lie in the same plane as the sheet.

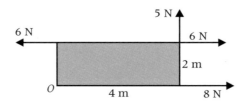

(a) Find the resultant moment of the four forces about the corner O.

(b) A force of magnitude 10 N is applied to the sheet, so that the total moment about O is zero. Draw a diagram to show where this force should be applied.

2 The diagram shows a light rod, of length 2 m, that is smoothly pivoted at O. A horizontal force of 200 N acts at B, which is at the centre of the rod. Find the magnitude of the force that acts at A if the rod is in equilibrium.

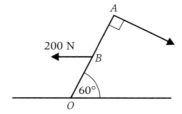

3 A uniform ladder has length 5 m and mass 20 kg. It leans against a smooth vertical wall, with one end on rough horizontal ground. The angle between the ladder and the ground is 70°. A child, of mass 40 kg, is at the top of the ladder. The coefficient of friction between the ladder and the ground is μ.

(a) Draw a diagram to show the forces acting on the ladder.

(b) Show that the magnitude of the friction force acting on the base of the ladder is approximately 178 N.

(c) Calculate the magnitude of the normal reaction force acting on the base of the ladder.

(d) Find the minimum value of μ for the ladder to remain at rest.
[A]

4 A loft door *OA* of weight 100 N is propped open at 50° to the horizontal by a strut *AB*. The door is hinged at *O*; *OA* = *OB* = 1 m. Assuming that the mass of the strut can be neglected compared to the mass of the door and that the weight of the door acts through the midpoint of *OA*, find:

   (a) the force in the strut,

   (b) the reaction at the hinge.

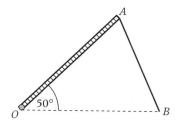

5 A ladder, of length 3 m and mass 20 kg, leans against a smooth, vertical wall so that the angle between the horizontal ground and the ladder is 60°. Find the magnitude of the friction and normal reaction forces that act on the ladder, if it is in equilibrium.

6 A uniform rod of length 2 m and mass 5 kg is connected to a vertical wall by a smooth hinge at *A* and a wire *CB* as shown. If a 10 kg mass is attached to *D*, find:

   (a) the tension in the wire,

   (b) the magnitude of the reaction at the hinge *A*.

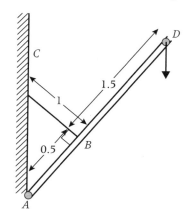

7 The foot of a uniform ladder of mass m rests on rough horizontal ground and the top of the ladder rests against a smooth vertical wall. When a man of mass $4m$ stands at the top of the ladder the system is in equilibrium with the ladder inclined at 60° to the horizontal. Find the coefficient of friction between the ladder and the ground.

14 Moments and equilibrium

8 The diagram shows a man holding a rope attached to one end of a uniform metal rod, of mass 200 kg, that is freely pivoted at A. A 120 kg mass is attached to the other end of the rod by another rope. Initially the rod is horizontal.

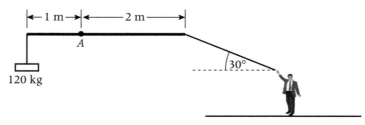

(a) Draw a labelled diagram to show the forces acting on the rod.

(b) Find the tension in the rope.

9 The diagram shows a man of mass 70 kg at rest while abseiling down a vertical cliff. Assume that the rope is attached to the man at his centre of mass. *You should model the man as a uniform rod and assume that he is not holding the rope.*

(a) Draw a diagram to show the forces acting on the man.

(b) Find the minimum value of the coefficient of friction between his feet and the cliff if he remains at rest.

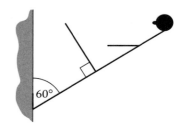

10 A uniform ladder, of mass m, leans against a vertical wall with its base on horizontal ground. The length of the ladder is 6 m. Assume that the wall is smooth and that the ground is rough, with the coefficient of friction between the ladder and the ground equal to 0.5.

(a) If the angle between the ladder and the ground is θ, show that the ladder remains at rest if θ is greater than or equal to 45°.

A person of mass M climbs the ladder.

(b) Show that when the person is at a distance x m from the bottom of the ladder

$$\tan \theta \geq \frac{3m + Mx}{3(m + M)}$$

if the ladder is to remain at rest.

(c) How far up the ladder can the person climb if $\theta = 45°$?

(d) Now assume that $\tan \theta = 2$.

  (i) Show that $x \leq 6 + \dfrac{3m}{M}$, if the ladder remains at rest.

  (ii) Use this result to make a prediction about the mass of a person who can reach the top of the ladder. [A]

Key point summary

Formulae to learn:

> Moment of a force = Force × Perpendicular distance
>
> Moment of the force about $O = Fd$
>
> Moment of force about $O = Fd \sin \theta$.

- The moment of a force about a point is the magnitude of the force multiplied by the perpendicular from the point to the force.   p2

- Moments can have a clockwise or anticlockwise turning effect.   p2

- The resultant force and moment must both be zero for a rigid body to be in equilibrium.   p4

Test yourself

What to review

1 A uniform metal beam, of mass 6 kg and length 2 m, rests in a horizontal position on two supports that are at a distance of 40 cm from each end of the beam. A 1.2 kg mass is placed at one end of the beam.

    Section 1.3

   (a) Find the magnitude of the reaction forces acting on the beam.

   (b) What is the greatest mass that could be placed at the other end of the beam, if it is to remain in equilibrium?

2 A ladder, of mass 20 kg and length 5 m, has its base on rough, horizontal ground and rests against a smooth vertical wall. The coefficient of friction between the ground and the ladder is 0.6. The angle between the ladder and the ground is θ.

    Section 1.4

   (a) Find the magnitude of the forces acting on the base of the ladder in terms of g and θ.

   (b) Find the smallest value of θ for which the ladder will remain at rest.

Test yourself ANSWERS

1 (a) 25.5 N, 45.1 N;

   (b) 13.8 kg.

2 (a) $R = 20g$, $F = \dfrac{10g}{\tan \theta}$;

   (b) 39.8°.

CHAPTER 2.
Centres of mass

Learning objectives

After studying this chapter you should be able to:
- find the centre of mass of a system of particles or of a composite body
- know that, when a body is suspended in equilibrium from a point, the centre of mass is directly below the point of suspension.

2.1 Centre of mass of a system of particles

In this section we will consider how to find the centre of mass of a system of particles. To illustrate the principle that we will use, imagine a light rod which has a different sized mass fixed to each end. There will be a point on the rod at which it can be balanced. This point is called the centre of mass.

Balance point

The position of the centre of mass can be found using moments.

The diagram shows a rod with two particles of masses, m_1 and m_2, which are at distances x_1 and x_2 from the left-hand end. Assume that the rod can be balanced by a single force, R, acting upwards at a distance $\bar{x}$ from the left-hand end.

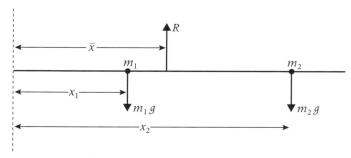

For the rod to balance the upward force must balance the two downwards forces, so that $R = m_1 g + m_2 g$.

Now taking moments about the left-hand end of the rod gives:

$$R\bar{x} - m_1 g x_1 - m_2 g x_2 = 0$$

$$\bar{x} = \frac{m_1 g x_1 + m_2 g x_2}{R}$$

$$= \frac{m_1 g x_1 + m_2 g x_2}{m_1 g + m_2 g}$$

$$= \frac{m_1 x_1 + m_2 x_2}{m_1 + m_2}$$

> This principle can be extended to find the centre of mass of any number of particles, using the general result:
>
> $$\bar{x} = \frac{\sum_{i=1}^{n} m_i x_i}{\sum_{i=1}^{n} m_i}$$

The worked examples below illustrate how this can be applied.

Worked example 2.1

Particles of masses 5 kg, 3 kg and 2 kg are fixed to a light rod of length 1.2 m. The 3 kg mass is in the centre and the others are at the ends of the rod. Find the distance of the centre of mass from the 5 kg mass.

Solution

The diagram shows the positions of the masses on the rod.

```
   5 kg          3 kg          2 kg
    •─── 0.6 m ───•─── 0.6 m ───•
```

Using the formula $\bar{x} = \dfrac{\sum_{i=1}^{n} m_i x_i}{\sum_{i=1}^{n} m_i}$

$$\bar{x} = \frac{5 \times 0 + 3 \times 0.6 + 2 \times 1.2}{5 + 3 + 2}$$

$$= \frac{4.2}{10}$$

$$= 0.42 \text{ m}$$

Working in two dimensions

For a system of particles in two dimensions the result above can still be used. The diagram shows a system of four particles in two dimensions.

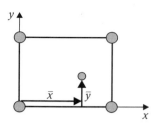

The position of the centre of mass relative to the bottom left-hand corner of the system has been shown. The coordinates of this point relative to the corner are $\bar{x}$ and $\bar{y}$. These are calculated using

$$\bar{x} = \frac{\sum_{i=1}^{n} m_i x_i}{\sum_{i=1}^{n} m_i} \quad \text{and} \quad \bar{y} = \frac{\sum_{i=1}^{n} m_i y_i}{\sum_{i=1}^{n} m_i}$$

where x_i and y_i are the coordinates of the particles.

Worked example 2.2

Particles of mass 4 kg, 5 kg, 8 kg and 3 kg are fixed to the corners of a light square framework, with sides of length 0.6 m. The diagram shows the positions of the masses.

Find the coordinates of the centre of mass of the system, with respect to the 3 kg mass.

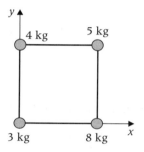

Solution

Using the formula $\bar{x} = \dfrac{\sum_{i=1}^{n} m_i x_i}{\sum_{i=1}^{n} m_i}$ gives:

$$\bar{x} = \frac{0 \times 3 + 0 \times 4 + 0.6 \times 8 + 0.6 \times 5}{3 + 4 + 8 + 5}$$

$$= \frac{7.8}{20}$$

$$= 0.39$$

Similarly using the formula $\bar{y} = \dfrac{\sum_{i=1}^{n} m_i y_i}{\sum_{i=1}^{n} m_i}$ gives:

$$\bar{y} = \frac{0 \times 3 + 0.6 \times 4 + 0 \times 8 + 0.6 \times 5}{3 + 4 + 8 + 5}$$

$$= \frac{5.4}{20}$$

$$= 0.27$$

So the coordinates relative to the 3 kg mass are (0.39, 0.27).

Worked example 2.3

A light rectangular framework has sides of length 2 m and 1 m. Particles of mass 2 kg, 4 kg, M kg and m kg are fixed to the corners of the framework as shown in the diagram.

The coordinates of the centre of mass relative to the bottom left-hand corner are (1.6, 0.3). Find M and m.

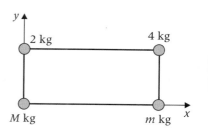

Solution

First consider the x-coordinate of the centre of mass.

$$1.6 = \frac{0 \times M + 0 \times 2 + 2 \times 4 + 2 \times m}{2 + 4 + m + M}$$

$1.6(6 + m + M) = 8 + 2m$

$9.6 + 1.6m + 1.6M = 8 + 2m$

$1.6 = 0.4m - 1.6M$

$4 = m - 4M$ \hfill [1]

Now consider the y-coordinate:

$$0.3 = \frac{0 \times M + 1 \times 2 + 1 \times 4 + 0 \times m}{2 + 4 + m + M}$$

$0.3(6 + m + M) = 6$

$1.8 + 0.3m + 0.3M = 6$

$4.2 = 0.3m + 0.3M$

$14 = m + M$ \hfill [2]

We now have a pair of simultaneous equations. Eliminating m from these equations, by subtracting [1] from [2], gives:

$10 = 5M$

$M = 2$ kg

Then substituting this value into equation [2] gives:

$14 = m + 2$

$m = 12$ kg

Worked example 2.4

A rod has mass 5 kg and length 60 cm. A particle of mass 12 kg is fixed at one end and a particle of mass 8 kg is fixed at the other. Find the distance of the centre of mass from the 12 kg mass.

Solution

Model the rod as a particle of mass 5 kg at its centre, as shown in the diagram.

Then using the formula $\bar{x} = \dfrac{\sum_{i=1}^{n} m_i x_i}{\sum_{i=1}^{n} m_i}$ gives:

$$\bar{x} = \frac{12 \times 0 + 5 \times 0.3 + 8 \times 0.6}{12 + 5 + 8}$$

$$= \frac{6.3}{25}$$

$$= 0.252 \text{ m or } 25.2 \text{ cm}$$

EXERCISE 2A

1. Particles of mass 3 kg and 6 kg are placed at opposite ends of a light rod of length 1.8 m. Find the distance of the centre of mass from the 3 kg mass.

2. A 5 kg particle is placed at the centre of a light rod of length 0.8 m. A particle of mass 2 kg is fixed to one end of the rod and a particle of mass 1 kg is fixed to the other end. Find the distance of the centre of mass from the 1 kg mass.

3. Four particles of mass 5 kg, 4 kg, 3 kg and 6 kg lie on the same horizontal line as shown. How far is their centre of mass from the 5 kg particle?

4. Three particles A, B and C, of masses 5 kg, 3 kg and 4 kg, respectively, lie on the same horizontal line as shown. Find the distance of their centre of mass from A.

   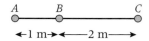

   The particle A is removed and replaced by a new particle. What is the greatest value of its mass if the centre of mass of the three particles is to lie in BC?

5. A rod, of length 50 cm, has three masses attached to it. There is an 8 kg mass at one end and a 7 kg mass at the other end. A 5 kg mass is also attached at another point on the rod. The centre of mass of the rod is 20 cm from the 8 kg mass.

   Find how far the 5 kg mass is from the 8 kg mass, if:

   (a) the rod is light,

   (b) the rod is uniform and has a mass of 5 kg.

6 For each light framework shown below find the coordinates of the centre of mass of the particles attached to the framework, with respect to the corner marked O.

(a)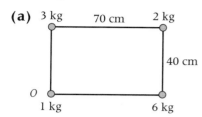

(b)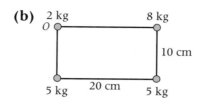

(c)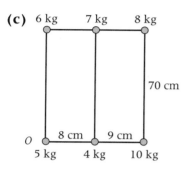

(d)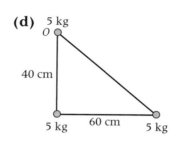

7 Particles of mass m, $2m$, $3m$ and $4m$ are attached to the corners of a square framework with sides of length l. The framework is shown in the diagram. Find the position of the centre of mass with respect of the corner marked O, if:

(a) the framework is light,

(b) the framework is made up of 4 rods of mass m.

8 The diagram shows a light rectangular framework of height 40 cm and width 20 cm. Particles are attached to the corners of the framework as shown in the diagram. The centre of mass is at a height of 16 cm and at a horizontal distance of 14 cm from the left-hand edge of the framework.

Find m and M.

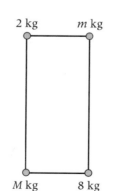

9 A light, triangular framework has sides of length 30 cm, 40 cm and 50 cm. Particles of mass 3 kg, 2 kg and m kg are attached to it as shown in the diagram.

(a) Find m if the centre of mass is to be at a height of 10 cm above the base.

(b) Find the horizontal distance of the centre of mass from the 3 kg mass.

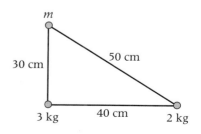

10 Particles are fixed to the ends of a light cross shape as shown in the diagram. The cross is made from two light rods of length 40 cm. The rods are joined at their centres, so that they are perpendicular. Find the distance of the centre of mass of the system from the centre of the cross, *O*.

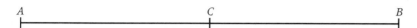

11 A uniform rod, *AB*, has length 4 m, and *C* is the midpoint of *AB*.

```
A                     C                     B
├─────────────────────┼─────────────────────┤
```

Three particles are attached to the rod: one of mass 7 kg at *A*; one of mass 12 kg at *C*; and one of mass 11 kg at *B*.

The mass of the uniform rod *AB* is 10 kg.

(a) Find the distance of the centre of mass of the system of rod and particles from *A*.

(b) The rod rests in a horizontal position supported at *A* and *B*.

Find the magnitudes of the vertical forces exerted by the supports on the rod at *A* and *B*. [A]

2.2 Centre of mass of a composite body

In this section we will find the centres of mass of composite bodies. By a composite body we mean something formed when two or more parts are joined together, for example a disc attached to the end of a rod. We will also consider shapes that have had holes cut in them. Some simple composite bodies are illustrated.

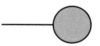

A disc attached to a rod

To find the centre of mass of a composite body we break it down into two or more shapes and find the centre of mass of each of these shapes. This is very often obvious because the shapes will be simple ones like rectangles, squares or circles. Then the centre of mass can be found by using the same approach that we have used for particles in the last section, by assuming that there is a particle at the centre of mass of each of the shapes that are being considered.

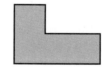

A body formed by joining two rectangles

You will find the word lamina is often used in this context. A lamina is a thin sheet of material. The thickness of the lamina is negligible. Also when working with composite body problems it is important to check that the bodies are uniform, that is they are made entirely from the same material.

A circle with a hole cut in it

Worked example 2.5

A disc of mass 4 kg and radius 18 cm is attached to the end of a rod of mass 5 kg and length 180 cm. Find the distance of the centre of mass from the base of the rod.

Solution

The diagram shows the centres of mass of the rod and the disc.

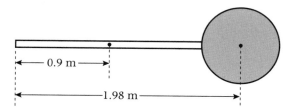

Using the formula $\bar{x} = \dfrac{\sum_{i=1}^{n} m_i x_i}{\sum_{i=1}^{n} m_i}$ gives:

$$\bar{x} = \frac{5 \times 0.9 + 4 \times 1.98}{4 + 5}$$

$$= \frac{12.42}{9}$$

$$= 1.38 \text{ m}$$

Centre of mass of a lamina

When working with a uniform lamina, as in the examples below, it is possible to work with areas instead of masses, because the area of each part will be proportional to its mass.

Worked example 2.6

The diagram shows a uniform lamina. Find the distance of the centre of mass from AB and the distance from AF.

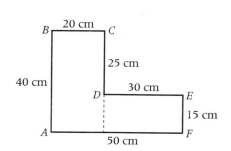

Solution

First find the total area of the lamina, which has been split into two parts on the diagram.

$$\text{Area} = 40 \times 20 + 30 \times 15$$

$$= 1250 \text{ cm}^2$$

To find the distance of the centre of mass from AB use the formula $\bar{x} = \dfrac{\sum_{i=1}^{n} m_i x_i}{\sum_{i=1}^{n} m_i}$, but replacing the masses with the areas of each part.

$$\bar{x} = \frac{800 \times 10 + 450 \times 35}{1250}$$

$$= 19 \text{ cm}$$

Similarly to find the distance from AF, use the formula

$$\bar{y} = \frac{\sum_{i=1}^{n} m_i y_i}{\sum_{i=1}^{n} m_i}$$

$$\bar{y} = \frac{800 \times 20 + 450 \times 7.5}{1250}$$

$$= 15.5 \text{ cm}$$

The position of a suspended lamina or body

When a lamina or other body is suspended, it will hang, so that the centre of mass is directly below the point of suspension.

If you take a rectangle and suspend it from one corner, the rectangle will hang so that one diagonal is vertical, because the centre of mass is on the diagonal, as shown in the diagram.

The angle between the vertical and one side in this position can then be calculated.

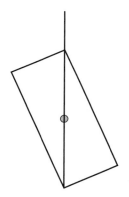

Worked example 2.7

The lamina from the last example is suspended from the corner B. The lamina remains at rest. Find the angle between the side AB and the vertical.

Solution

The diagram shows the lamina, as it would hang, with the centre of mass directly below B.

To find the angle α, as required, note that:

$$\tan \alpha = \frac{\bar{x}}{40 - \bar{y}}$$

$$= \frac{19}{40 - 15.5}$$

$$\alpha = 37.8°$$

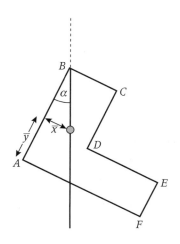

Worked example 2.8

The diagram shows a uniform rectangular lamina that has had a hole cut in it. The centre of mass of the lamina is a distance x from AD and a distance y from AB. Find x and y. The lamina is suspended from the corner A. Find the angle between AB and the vertical.

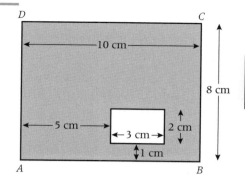

Solution

Consider the table below:

	Area	Distance of C of M from AD	Distance of C of M from AB
Large rectangle	80 cm²	5 cm	4 cm
Small rectangle	6 cm²	6.5 cm	2 cm
Lamina	74 cm²	x	y

Considering the large rectangle as being made up of the lamina and a small rectangle allows us to formulate the equation below, where x is the distance of the centre of mass from AD:

$$80 \times 5 = 74x + 6 \times 6.5$$
$$400 = 74x + 39$$
$$x = \frac{400 - 39}{74}$$
$$= \frac{361}{74} = 4.88 \text{ cm (to three significant figures)}$$

Similarly to find the distance, y, of the centre of mass from the side AB:

$$80 \times 4 = 74y + 6 \times 2$$
$$320 = 74y + 12$$
$$y = \frac{320 - 12}{74}$$
$$= \frac{308}{74} = 4.16 \text{ cm (to three significant figures)}$$

When hanging in equilibrium, the angle θ, between AB and the vertical is:

$$\theta = \tan^{-1}\left(\frac{y}{x}\right)$$
$$= \tan^{-1}\left(\frac{308}{74} \times \frac{74}{361}\right)$$
$$= 40.5°$$

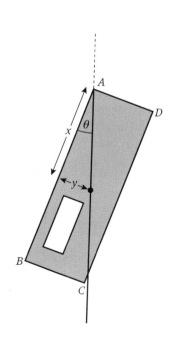

EXERCISE 2B

1. A uniform sphere of mass 2 kg and radius 20 cm is fixed to the end of a uniform rod of length 2 m and mass 4 kg. Find the distance of the centre of mass from the end of the rod.

2. A table tennis bat can be modelled as a uniform disc of radius 8 cm and mass 100 grams attached to the end of a uniform rod of mass 80 grams and length 5 cm. Find the distance between the bottom of the handle and the centre of mass.

3. A rectangular uniform lamina has sides of length 10 cm and 20 cm. A string is attached to one corner of the lamina and it is allowed to hang in equilibrium in a vertical plane. Find the angle between the vertical and the longer side of the rectangle.

4. A rectangular uniform lamina hangs in equilibrium from one corner. The dimensions of the rectangle are 45 cm and 60 cm. Find the angle between the vertical and the shorter sides of the lamina.

5. The diagram shows a uniform lamina.
   (a) Find the distance between the side AB and the centre of mass.
   (b) Find the distance between the side AF and the centre of mass.
   (c) If the lamina is suspended in equilibrium from the corner A, find the angle between AB and the vertical.
   (d) If the lamina is suspended in equilibrium from the corner B, find the angle between AB and the vertical.

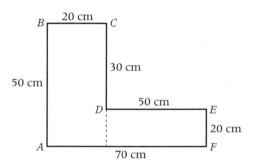

6. A uniform rectangle has sides of length 20 cm and 60 cm. A smaller rectangle, with sides of length 30 cm and 10 cm, is cut out of one corner of the rectangle, to form a lamina. The lamina is then allowed to hang from the opposite corner. When the lamina hangs in equilibrium, what is the angle between the vertical and its longest side?

7. The diagram shows a uniform lamina. Find the distance of the centre of mass from AB and from AF. The lamina is suspended from the corner A. Find the angle between the side AB and the vertical, when the lamina is at rest.

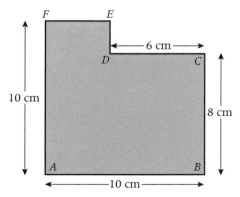

8 A letter F is made out of uniform card, by cutting out two rectangles of size 20 cm by 10 cm and one rectangle of size 40 cm by 10 cm. These are stuck together so that the rectangles overlap as shown in the diagram.

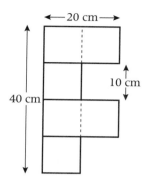

(a) If the F is suspended from the top left-hand corner, find the angle between the vertical and the 40 cm side.

(b) Describe how to suspend the letter, so that the top is horizontal.

9 A light, rectangular framework $ABCD$ has particles of mass 3 kg, 4 kg and 3 kg attached at A, B and C, respectively. When suspended from A the framework hangs with C directly below A.

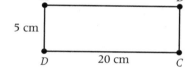

(a) Find the mass of the particle at D.

(b) The framework is suspended from a point X on AB, so that AB is horizontal. Find the distance of X from A.

10 The diagram shows a uniform lamina.

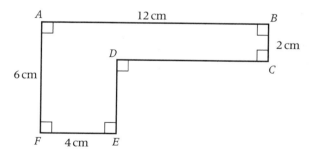

(a) Show that the centre of mass of the lamina is 2.2 cm from the side AB.

(b) Find the distance of the centre of mass of the lamina from the side AF.

(c) The lamina is suspended from the corner A and hangs in equilibrium. Find the angle between the side AF and the vertical.
[A]

11 The diagram shows a uniform lamina.

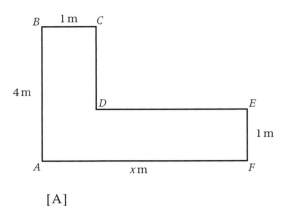

(a) For a particular lamina, $x = 7$.

  (i) Find the distance of the centre of mass of the lamina from the side AB.

  (ii) The lamina is suspended from the corner C. Find the angle between the side CD and the vertical.

(b) Another lamina is suspended from the corner C. Given that the side CD is vertical, find x. [A]

12

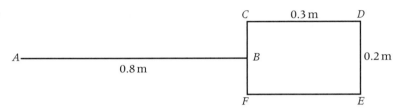

A garden spade can be modelled by a uniform rod, AB, together with a uniform rectangular plate CDEF. The plate is rigidly attached to the rod at B, which is the midpoint of CF, as shown in the diagram.

The rod is of length 0.8 m and mass 0.25 kg. The plate is such that $CD = FE = 0.3$ m, $CF = DE = 0.2$ m and its mass is 1 kg.

Find the distance of the centre of mass of the spade from A. [A]

13 The diagram shows a uniform lamina, which consists of two rectangles ABCD and DPQR.

The dimensions are such that:
$DR = PQ = CP = 12$ cm; $BC = QR = 8$ cm;
$AB = AR = 20$ cm.

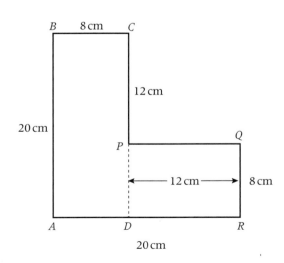

(a) Explain why the centre of mass of the lamina must lie on the line AP.

(b) Find the distance of the centre of mass of the lamina from AB.

(c) The lamina is freely suspended from B. Find, to the nearest degree, the angle that AB makes with the vertical through B. [A]

14 The diagram shows a uniform lamina which is to be used as part of a child's mobile. It consists of two rectangular parts *ABCD* and *PQRS*. The dimensions are such that $AB = CD = 2$ cm, $BC = AD = SR = PQ = 12$ cm and $RQ = SP = 6$ cm. The midpoint, *O*, of *CB* coincides with the midpoint of *SP*.

(a) Find the distance of the centre of mass of the lamina from *RQ*.

(b) The lamina is suspended from *Q*. Find the angle that *RQ* makes with the vertical through *Q*. [A]

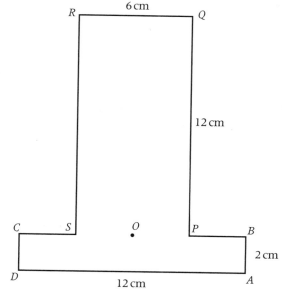

15 A letter P is formed by bending a uniform steel rod into the shape shown below, in which *ABCD* is a rectangle.

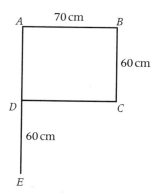

(a) Find the distance of the centre of mass of the letter from the side:

(i) *AE*,   (ii) *AB*.

The letter is to be suspended from a point *F* on the side *AB*. The point *F* is a distance *x* cm from *A*.

(b) State the value of *x* if the side *AB* is to be horizontal.

(c) Find the value of *x* if the side *AB* is to be at an angle of 5° to the horizontal, with *A* higher than *B*. [A]

16 A circular hole, of radius 4 cm, is cut in a uniform, circular disc of radius 10 cm.

Find the distance of the centre of mass from the point A.

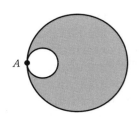

17 A disc, of radius 20 cm, has centre O. A hole of radius 5 cm is drilled in the disc, so that the centre of the hole is 5 cm from O and lies on the line BC, that passes through O. The disc is suspended from the point A, so that the line BC is horizontal. Find the angle BOA.

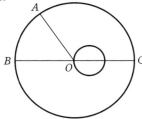

18 A square piece of thin uniform metal has sides of length 5 cm and a rectangular hole cut in it. The diagram shows the position of the hole.

(a) If x cm is the distance of the centre of mass from the side AD, show that $x = \dfrac{119}{46}$.

(b) If y cm is the distance of the centre of mass from the side AB, find y.

(c) The metal sheet is hung from a smooth peg that passes through the rectangular hole. In equilibrium the sheet is at rest, in a vertical plane, with the peg in the corner of the rectangle closest to D. Find the angle between the side AD and the vertical. [A]

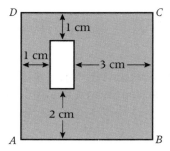

Key point summary

Formulae to learn: *p17*

$$\bar{x} = \dfrac{\sum_{i=1}^{n} m_i x_i}{\sum_{i=1}^{n} m_i} \text{ and } \bar{y} = \dfrac{\sum_{i=1}^{n} m_i y_i}{\sum_{i=1}^{n} m_i}$$

- The centre of mass of a system of particles or of a composite body can be found using moments. *p16*

- When a body is suspended in equilibrium from a point, the centre of mass is directly below the point of suspension. *p24*

Centres of mass

Test yourself

What to review

1 The diagram shows four particles attached to the corners of a light rectangular framework.

Section 2.1

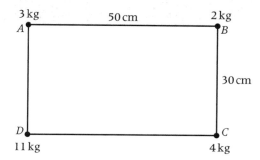

(a) Find the distance of the centre of mass from

(i) AB, (ii) AD.

(b) Calculate the angle between AB and the vertical when the framework is suspended in equilibrium from A.

2 The diagram shows a uniform lamina $ABCDEF$.

Section 2.2

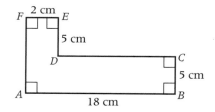

(a) Show that the centre of mass of the lamina is 8.2 cm from AF.

(b) Find the distance of the centre of mass from AB.

(c) If the lamina is suspended from the corner A, find the angle between AB and the vertical.

Test yourself ANSWERS

1 (a) (i) 22.5 cm, (ii) 15 cm. **(b)** 56.3°.

2 (b) 3 cm. **(c)** 20.1°.

CHAPTER 3
Energy

Learning objectives

After studying this chapter you should be able to:
- calculate kinetic energy
- calculate work done by a constant force
- calculate gravitational potential energy
- calculate elastic potential energy
- be able to use power.

3.1 Kinetic energy

Every moving body has kinetic energy. The greater the mass and the greater the speed, the greater the kinetic energy.

> The kinetic energy of a body is defined as $\frac{1}{2}mv^2$, where m is the mass of the body and v its speed.

The units of energy are joules (J).

Worked example 3.1

A car has mass 1100 kg. At the bottom of a hill it is travelling at 30 m s^{-1} and loses speed as it travels up the hill. At the top of the hill its speed is 22 m s^{-1}. Calculate the amount of kinetic energy lost as the car drove up the hill.

Solution

At the bottom of the hill:

$$\text{Kinetic energy} = \tfrac{1}{2} \times 1100 \times 30^2$$
$$= 495\,000 \text{ J}$$

At the top of the hill:

$$\text{Kinetic energy} = \tfrac{1}{2} \times 1100 \times 22^2$$
$$= 266\,200 \text{ J}$$

Now the amount of kinetic energy that has been lost can be calculated.

$$\text{Kinetic energy lost} = 495\,000 - 266\,200$$
$$= 228\,800 \text{ J}$$

EXERCISE 3A

1. Calculate the kinetic energy of a ball, of mass 150 grams, travelling at 8 m s^{-1}.

2. Calculate the kinetic energy of a train, of mass 30 000 tonnes, travelling at 50 m s^{-1}.

3. A ball has a mass of 200 grams. It is thrown so that its initial speed is 12 m s^{-1} and during its flight it has a minimum speed of 6 m s^{-1}. Calculate the minimum and maximum values of the kinetic energy of the ball.

4. A light aeroplane has a mass of 1500 kg. When it lands it is travelling at 80 m s^{-1} and at the end of the runway its speed has been reduced to 10 m s^{-1}. Calculate how much kinetic energy has been lost.

5. A stone, of mass 50 grams, is dropped from the top of a cliff at a height of 40 m.
   (a) Assume that no resistance forces act on the stone. Use a constant acceleration equation to calculate its speed at the bottom of the cliff.
   (b) How much kinetic energy does the stone gain as it falls?

6. A cycle and cyclist have mass 70 kg. The cyclist freewheels from rest down a slope, accelerating at 0.5 ms^{-2}. The initial speed of the cyclist is 3 m s^{-1}.
   (a) Calculate the speed of the cyclist after he has travelled 50 m, by using the equation $v^2 = u^2 + 2as$.
   (b) Calculate the increase in the kinetic energy of the cyclist.

3.2 Work and energy

As a stone falls its kinetic energy increases. As you start to pedal a cycle your kinetic energy increases. In the first of these cases gravity is the force that causes a change in kinetic energy. In the second the cyclist exerts a force. In this section you will examine the relationship between the change in the kinetic energy of a body and the forces that act on it.

If a constant force of magnitude F acts on a body of mass m, it will produce an acceleration of $\dfrac{F}{m}$, this can be substituted into the constant acceleration equation $v^2 = u^2 + 2as$, to give

$$v^2 = u^2 + 2 \times \dfrac{F}{m} \times s$$

or

$$\tfrac{1}{2}mv^2 - \tfrac{1}{2}mu^2 = Fs$$

This equation can be expressed as

Change in kinetic energy $= Fs$

The quantity *Fs* is referred to as the work done by the force. It is this work that determines the change in the kinetic energy of the body that the force causes.

Worked example 3.2

A ball, of mass 0.4 kg, is released from rest and allowed to fall 3 m.

(a) Find the work done by gravity as the ball falls.

(b) State the gain in kinetic energy of the ball.

(c) Calculate the speed of the ball when it has fallen 3 m.

Solution

(a) The work done is calculated using *Fs*. In this case

$$F = mg$$
$$= 0.4 \times 9.8$$
$$= 3.92 \text{ N}$$

The work done by gravity can be calculated.

$$\text{Work done} = Fs$$
$$= 3.92 \times 3$$
$$= 11.76 \text{ J}$$

(b) As the gain in kinetic energy is equal to the work done, there is a gain in kinetic energy of 11.76 J.

(c) The kinetic energy of the ball is 11.76 J, so the speed can be calculated as below;

$$11.76 = \tfrac{1}{2} \times 0.4 v^2$$
$$v^2 = 58.8$$
$$v = \sqrt{58.8}$$
$$= 7.67 \text{ m s}^{-1} \text{ (to 3 sf)}$$

Worked example 3.3

A box is initially at rest on a smooth horizontal surface. The mass of the box is 5 kg. A horizontal force of magnitude 8 N acts on the box as it slides 6 m.

(a) Find the work done by the force.

(b) Find the speed of the box when it has travelled 6 m.

Solution

(a) The work done is calculated using Fs.

$$\text{Work done} = 8 \times 6$$
$$= 48 \text{ J}$$

(b) Using the fact that the work done is equal to the change in kinetic energy gives

$$48 = \tfrac{1}{2} \times 5v^2$$
$$v = \sqrt{19.2}$$
$$= 4.38 \text{ m s}^{-1} \text{ (to 3 sf)}$$

In most situations more than one force will act. Some forces may act in the direction of motion, as in the previous examples, but often they will act in the opposite direction to the motion. These types of forces will include resistance and friction forces. A force that acts in the opposite direction to the motion will do a negative amount of work. For example the work done by a friction force of magnitude 80 N acting on a body that moves 5 m would be −400 J. Often we would say that the work done against friction is 400 J.

Worked example 3.4

A car, of mass 1250 kg, is subject to a forward force of magnitude 2000 N and a resistance force of magnitude 500 N. The car moves 200 m.

(a) Find the work done by each of the forces acting on the car.

(b) If the car is initially moving at 5 m s^{-1}, find the final speed of the car.

Solution

(a) The work done by the 2000 N force is $2000 \times 200 = 400\,000$ J.

The work done by the 500 N force is $-500 \times 200 = -100\,000$ J. We might say that the work done against friction is 100 000 J.
Total work done = 400 000 − 100 000 = 300 000 J

Alternatively note that the resultant force is 1500 N and the work done will be $1500 \times 200 = 300\,000$ J.

(b) The change in kinetic energy is 300 000 J. So the final speed can be calculated

$$300\,000 = \tfrac{1}{2} \times 1250 \times v^2 - \tfrac{1}{2} \times 1250 \times 5^2$$
$$625 v^2 = 315\,625$$
$$v^2 = 505$$
$$v = \sqrt{505}$$
$$= 22.5 \text{ m s}^{-1} \text{ (to 3 sf)}$$

Worked example 3.5

A ball, of mass 0.3 kg, is moving at 8 m s^{-1} when it enters a tank of water. It hits the bottom of the tank travelling at 2 m s^{-1}. The depth of water in the tank is 1.2 m. Assume that a constant resistance force acts on the ball as it moves through the water.

(a) Calculate the change in the kinetic energy of the ball.

(b) Find the magnitude of the resistance force that acts on the ball.

Solution

(a) Change in kinetic energy $= \frac{1}{2} \times 0.3 \times 2^2 - \frac{1}{2} \times 0.3 \times 8^2$
$= -9$ J

(b) First consider the work done by each of the forces.

Work done by gravity $= 0.3 \times 9.8 \times 1.2$
$= 3.528$ J

If the resistance force has magnitude R, then the work done by this force will be

$-R \times 1.2 = -1.2R$

and so the total work done is

$3.528 - 1.2R$.

Now we can find R by using

Change in kinetic energy = work done

$-9 = 3.528 - 1.2R$

$R = \dfrac{3.528 + 9}{1.2}$

$= 10.44$ N

EXERCISE 3B

1 A force of magnitude 800 N acts on a car, of mass 1000 kg, as it moves 400 m on a horizontal surface. Assume that no other forces act on the car.

(a) Calculate the work done by the force that acts on the car.

(b) Find the final kinetic energy and speed of the car if it is initially

(i) at rest,

(ii) moving at 3 m s^{-1}.

2 A force acts horizontally on a package, of mass 4 kg, that is initially at rest on a smooth horizontal surface. After the package has moved 3 m its speed is 5 m s^{-1}.

   (a) Find the increase in the kinetic energy of the package.

   (b) How much work is done by the force that acts on the package?

   (c) Determine the magnitude of the force that acts on the package.

3 A brick, of mass 2 kg, is allowed to fall from rest at a height of 3.2 m. Find the kinetic energy and speed of the brick when it hits the ground;

   (a) assuming that no resistance forces act on the brick as it falls,

   (b) assuming that a constant resistance force of magnitude 10 N acts on the brick.

4 A rope is attached to a boat of mass 200 kg. The boat is pulled along a horizontal surface by the horizontal rope. The tension in the rope remains constant at 500 N. When the boat has moved 50 m its speed is 0.8 m s^{-1}.

   (a) Calculate the work done by the tension in the rope.

   (b) Calculate the final kinetic energy of the boat.

   (c) Find the work done against the resistance forces acting on the boat.

   (d) Find the magnitude of the resistance force if it is assumed to be constant.

5 A ball, of mass 200 grams, is dropped from a height of 1 m. The ball is initially at rest. It hits the ground and rebounds at $\frac{3}{4}$ of the speed with which it hit the ground.

   (a) Calculate the work done by gravity as the ball falls.

   (b) Calculate the speed of the ball when it hits the ground.

   (c) Find the maximum height of the ball after it bounces.

6 A forward force of magnitude 2500 N acts on a car, of mass 1250 kg, as it moves 500 m. The initial speed of the car was 12 m s^{-1}. All the motion takes place along a straight line.

   (a) If no resistance forces act on the car find its final kinetic energy and speed.

   (b) If the final speed of the car is 30 m s^{-1}, find the work done against the resistance forces and the average magnitude of the resistance force.

7 As a car, of mass 1200 kg, skids 25 m, on a horizontal surface, its speed is reduced from 30 m s^{-1} to 20 m s^{-1}.
   (a) Find the energy lost by the car as it skids.
   (b) If the coefficient of friction between the car and the road is 0.8, find the work done by the friction force.
   (c) Find the work done by the air resistance force that acts on the car.
   (d) Assuming that the air resistance force is constant, find out how much further the car travels before it stops.
   (e) Criticise the assumptions that you have used to get your answer to (d) and suggest how using a more realistic model would change your answer.

8 A diver, of mass 60 kg, dives from a diving board at a height of 4 m. She hits the water travelling at a speed of 8 m s^{-1} and descends to a depth of 2 m in the diving pool. Model the diver as a particle, that is initially at rest.
   (a) Find the work done against air resistance before the diver hits the water and the average magnitude of the air resistance force.
   (b) Find the average magnitude of the force exerted by the water on the diver as she is brought to rest in the diving pool.

9 A ball of mass 500 grams is released from rest at a height of 3 m. It hits the ground and travels down 6 cm into soft mud before stopping. Assume no air resistance acts on the ball as it falls.
   (a) Find the average force that the mud exerts on the ball while it is slowing down.
   (b) A different ball of mass 200 grams hits the mud travelling at 6 m s^{-1}. Assume that the mud exerts the same force on the ball and find the distance that it travels into the mud before stopping.

10 A bullet fired into an earth bank at an unknown speed penetrates to a distance of 2 m. An identical bullet fired at a speed of 120 m s^{-1} travels 3 m into the bank. If the mass of this type of bullet is 20 grams, find the average force exerted on the bullets by the earth, and the speed at which the first bullet was fired.

3.3 Forces at angles

If a force acts at an angle to the direction of motion to the body that it acts on, then we must use the component of the force in the direction of motion when calculating the work done.

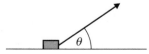

If the force shown in the diagram acts as the body moves a distance d, then the work done is $Fd \cos \theta$.

Worked example 3.6

A force, of magnitude 60 N, acts at an angle of 30° above the horizontal on a sack, of mass 50 kg, that is initially at rest on a horizontal surface. Find the work done by the force and the final speed of the sack which moves 5 m if:

(a) the surface is smooth,

(b) the coefficient of friction between the sack and the surface is 0.1.

Solution

(a) The diagram shows the 60 N force which acts on the sack. The work done can be calculated using $Fd \cos \theta$.

$$\text{Work done} = 60 \times 5 \times \cos 30°$$
$$= 300 \cos 30°$$

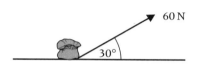

The work done is equal to the gain in kinetic energy, so

$$300 \cos 30° = \tfrac{1}{2} \times 50v^2$$
$$v^2 = 12 \cos 30°$$
$$v = 3.22 \text{ m s}^{-1} \text{ (to 3 sf)}$$

(b) The diagram shows all the forces acting on the sack. The work done by the force will be the same, but the final kinetic energy will be less due to the friction force. First we must calculate the magnitude of the friction force.

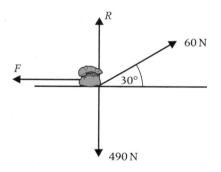

Resolving vertically to find the normal reaction, R, we have

$$R + 60 \sin 30° = 50 \times 9.8$$
$$R = 460 \text{ N}$$

Then the magnitude of the friction force can now be found using $F = \mu R$, which in this case gives

$$F = 0.1 \times 460$$
$$= 46 \text{ N}$$

The work done by the friction force will be

$$-46 \times 5 = -230 \text{ J}$$

The total work done is then $300 \cos 30° - 230$. The speed can now be found using work done equals change in kinetic energy.

$$300 \cos 30° - 230 = \tfrac{1}{2} \times 50v^2$$
$$v = 1.09 \text{ m s}^{-1} \text{ (to 3 sf)}$$

Worked example 3.7

A particle, of mass 5 kg, is initially at rest. It slides 4 m down a slope at 30°. Assume that the slope is smooth and that there is no air resistance.

(a) Find the work done by gravity as the particle slides down the slope.

(b) Find the speed of the particle, when it has travelled the 4 m.

Solution

(a) The work done can be calculated using $Fd \cos \theta$.

$$\text{Work done} = 5 \times 9.8 \times 4 \cos 60°$$
$$= 98 \text{ J}$$

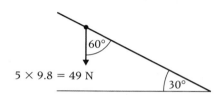

(b) Using work done equals change in kinetic energy gives

$$98 = \tfrac{1}{2} \times 5v^2$$
$$v = \sqrt{39.2}$$
$$= 6.26 \text{ m s}^{-1} \text{ (to 3 sf)}$$

Gravitational potential energy

Consider a particle of mass m that falls vertically a distance h, in the absence of any resistance forces. The work done by gravity will be given by mgh.

Also consider a second particle of the same mass that slides down the smooth slope shown in the diagram.

The particle will slide a distance of $\dfrac{h}{\sin \theta}$.

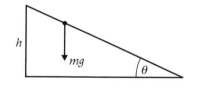

So the work done by gravity is

$$mg \times \frac{h}{\sin \theta} \times \cos(90 - \theta) = mg \times \frac{h}{\sin \theta} \times \sin \theta = mgh$$

Note that in both of these examples the work done by gravity is the same and depends on the initial height and not the route taken. A curved surface would have also produced the same result. The important feature of this result is that there are no resistance or friction forces present.

This quantity mgh is often referred to as the gravitational potential energy of the body and often this is abbreviated to potential energy or PE. If a body is allowed to fall, swing or slide the potential energy will be converted to kinetic energy. Similarly as a body rises its kinetic energy will be converted to potential energy. This gives a useful way of approaching some problems, because the total energy will remain constant if no resistance or friction forces act.

> Gravitational potential energy = mgh

Worked example 3.8

A soldier, of mass 72 kg, on a training exercise is running at a speed of 5 m s^{-1}. He grabs hold of a rope, of length 6 m, that is hanging vertically. He then swings on the rope. The angle between the rope and the vertical is θ.

(a) Calculate the initial kinetic energy of the soldier.

(b) Calculate the speed of the soldier when he has risen 0.5 m.

(c) Find the maximum height of the soldier and the angle between the rope and the vertical at this time.

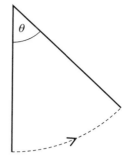

Solution

(a) The initial kinetic energy can be calculated from the information given

$$\text{KE} = \tfrac{1}{2} \times 72 \times 5^2$$
$$= 900 \text{ J}$$

(b) When the soldier has risen 0.5 m his potential energy can be calculated

$$\text{PE} = 72 \times 9.8 \times 0.5$$
$$= 352.8 \text{ J}$$

The remaining kinetic energy can then be calculated as

$$900 - 352.8 = 547.2$$

The speed of the soldier can now be found

$$547.2 = \tfrac{1}{2} \times 72v^2$$
$$v = \sqrt{15.2}$$
$$= 3.90 \text{ m s}^{-1} \text{ (to 3 sf)}$$

(c) At the soldier's highest point, all his initial kinetic energy will have been converted to potential energy. This gives the equation

$$72 \times 9.8h = 900$$
$$h = 1.28 \text{ m (to 3 sf)}$$

By considering the triangle in the diagram, which shows the maximum θ,

$$\cos \theta = \frac{6 - 1.28}{6}$$
$$\theta = 38° \text{ to the nearest degree.}$$

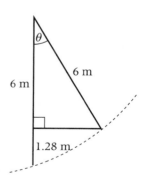

Worked example 3.9

A cyclist and cycle of combined mass 80 kg freewheel down a slope. They travel a distance of 100 m down the slope which is at an angle α to the horizontal, where $\sin \alpha = \tfrac{1}{10}$. The speed of the cyclist increases from 4 m s^{-1} to 8 m s^{-1}.

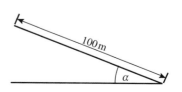

(a) Find the change in the total energy of the cycle and cyclist at the top and bottom of the slope.

(b) Find the work done against resistance forces while the cyclist travelled down the slope and the magnitude of the average resistance force on the cyclist.

(c) Find the speed of the cyclist when he has travelled 20 m down the slope.

Solution

(a) We must consider the gain in kinetic energy and the potential energy that is lost.

$$\text{Gain in KE} = \tfrac{1}{2} \times 80 \times 8^2 - \tfrac{1}{2} \times 80 \times 4^2$$
$$= 1920 \text{ J}$$

$$\text{Loss of PE} = 80 \times 9.8 \times 100 \sin \alpha$$
$$= 80 \times 9.8 \times 100 \times \tfrac{1}{10}$$
$$= 7840 \text{ J}$$

$$\text{Change in energy} = 1920 - 7840$$
$$= -5920 \text{ J}$$

So 5920 J of energy has been lost.

(b) The energy has been lost due to the work done against the resistance forces, so

$$\text{Work done against resistance forces} = 5920 \text{ J}$$

The average resistance force can be found by dividing the work done by the distance travelled.

$$\text{Average resistance force} = \frac{5920}{100} = 59.2 \text{ N}$$

(c) As the cycle and cyclist travel the 20 m, the resistance force will act. So first calculate the work done against these forces.

$$\text{Work done against resistance forces} = 20 \times 59.2 = 1184 \text{ J}$$

As the cycle and cyclist travel down the slope more potential energy is lost.

$$\text{Loss of PE} = 80 \times 9.8 \times 20 \times \tfrac{1}{10} = 1568 \text{ J}$$

The gain in kinetic energy will then be $1568 - 1184 = 384$ J. The initial KE is given by

$$\tfrac{1}{2} \times 80 \times 4^2 = 640 \text{ J}$$

The final kinetic energy will then be $640 + 384 = 1024$ J. Now the final speed can be found

$$\tfrac{1}{2} \times 80 v^2 = 1024 \text{ J}$$
$$v = \sqrt{25.6}$$
$$= 5.06 \text{ m s}^{-1} \text{ (to 3 sf)}$$

EXERCISE 3C

1 The diagram shows a curved slide with a drop of 5 m. A child, of mass 50 kg, sits at the top of the slide. He slides down. Assume that there are no resistance or friction forces acting on the slide. Calculate:

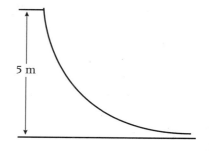

(a) the potential energy that the child would lose as he slides from the top to the bottom of the slide,

(b) the speed of the child at the bottom of the slide.

2 A stone, of mass 0.8 kg, is thrown over a cliff at a speed of 3 m s^{-1}. It hits the water at a speed of 12 m s^{-1}. Assume that there is no resistance to the motion of the stone.

(a) Find the initial potential energy of the stone.

(b) Find the height of the cliff.

3 A particle, of mass 3 kg, slides down a slope of length 20 m, which is inclined at an angle of 45° to the horizontal. At the top of the slope the particle has an initial speed of 4 m s^{-1}. Assume that the slope is smooth and that there is no air resistance.

(a) Find the potential energy lost by the particle as it slides down the slope.

(b) Find the kinetic energy and the speed of the particle at the bottom of the slope.

(c) If a constant friction force of magnitude 5 N acts on the particle as it slides, find the speed of the particle at the bottom of the slope.

4 A child, of mass 60 kg, swings on a rope of length 8 m. The rope is initially at an angle of 30° to the vertical. The child initially moves at 2 m s^{-1}.

(a) Find the potential energy that is lost as the child swings to her lowest point.

(b) Find the maximum kinetic energy and the maximum speed of the child.

(c) Find the maximum height of the child above her lowest position.

5 A ball of mass 300 grams is kicked so that it has an initial speed of 12 m s^{-1}. During its flight the speed of the ball has a minimum value of 4 m s^{-1}.

(a) Find the initial kinetic energy of the ball.

(b) Find the maximum potential energy of the ball.

(c) Find the maximum height of the ball.

6 A loop-the-loop roller coaster is shown in the diagram. At point A at the top of the first loop the roller coaster is moving at 8 m s^{-1}. The mass of the roller coaster carriage is 400 kg. Assume that no resistance forces act on the roller coaster. The diameter of the first loop is 5 m and the diameter of the second loop is 3 m.

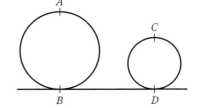

(a) Find the kinetic energy of the carriage at the bottom of the first loop.

(b) Find the kinetic energy of the carriage at the top of the second loop.

(c) Find the maximum speed of the roller coaster.

7 A sledge, of mass 12 kg, is pulled by a rope that is at an angle of 20° to the horizontal. The tension in the rope is a constant 80 N. The coefficient of friction between the sledge and the horizontal ground on which it moves is 0.2. Find the kinetic energy and the speed of the sledge when it has moved 5 m from rest.

8 A roller coaster, of mass 500 kg, is at the top of a slope and travelling at 4 m s^{-1}. As it travels down the slope its speed increases to 10 m s^{-1}. The length of the slope is 20 m and the top is 12 m higher than the bottom. At the bottom of the slope it travels on a horizontal section of track. Model the roller coaster as a particle that has a constant resistance force acting on it.

(a) Find the energy lost by the roller coaster as it moves down the slope.

(b) Find the magnitude of the resistance force on the roller coaster.

(c) Find how far the roller coaster travels along the horizontal section of the track before it comes to rest.

9 A car, of mass 1100 kg, is travelling down a hill, inclined at an angle of 5° to the horizontal. The driver brakes hard and skids 15 m as the car stops. The coefficient of friction between the tyres and the road is 0.7. Find the initial kinetic energy and speed of the car.

10 The diagram shows a slide that is in the shape of a semicircle of radius 5 m and that has centre O. The users slide down the inside of the slide and up the other side. They all start from rest at the top of the slide, at the point A. The mass of a user is 55 kg.

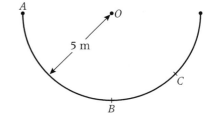

(a) If the slide is smooth find the maximum kinetic energy of a user and their maximum speed.

In fact the slide is not smooth. A simple model assumes that a constant resistance force acts on the users and the magnitude of this is 30 N.

(b) Calculate the speed of a user at the lowest point of the slide, marked B on the diagram.

11 A ball is projected vertically upwards, from ground level, with an initial speed of 18 m s^{-1}. The ball has a mass of 0.3 kg. Assume that the force of gravity is the only force acting on the ball after it is projected.

  (a) Calculate the initial kinetic energy of the ball.

  (b) By using conservation of energy, find the maximum height of the ball above ground level.

  (c) Find the kinetic energy and the speed of the ball when it is at a height of 2 m above ground level. [A]

12 A ball has mass 0.5 kg and is released from rest at a height of 6 m above ground level.

  (a) Assume that no resistance force acts on the ball as it falls.

  (i) Find the kinetic energy of the ball when it has fallen 3 m.

  (ii) Use an energy method to find the speed of the ball when it hits the ground.

  (b) Assume that a constant resistance force acts on the ball as it falls and that the ball hits the ground travelling at 2 m s^{-1}. Use an energy method to find the magnitude of the resistance force. [A]

13 The diagram shows a children's slide. The curved section AB is one quarter of a circle of radius 2.5 m, so that a child would be travelling horizontally at B. The horizontal section BC has length 5 m. A child of mass 45 kg uses the slide, starting from rest at A.

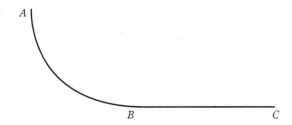

  (a) A simple model neglects friction and air resistance. Use this model to predict:

  (i) the kinetic energy of the child at the point C,

  (ii) her speed when she reaches the point C.

  A revised model assumes that a constant air resistance force of magnitude 20 N acts on the child as she slides from A to C.

  (b) Calculate the length of the slide between the points A and C. Hence find a revised prediction for the speed of the child at the point C.

A further revision to the model assumes that friction also acts on the child on the section BC and that the coefficient of friction between the child and the slide is 0.5.

(c) Find the distance that the child slides beyond the point B. [A]

14 A soldier, of mass 80 kg, swings on a rope of length 80 m. He is to be modelled as a particle that describes a circular arc from A, through B to C. The path is shown in the following diagram.

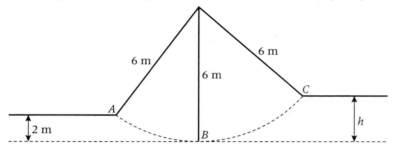

The point A is 2 m higher than B and C is h m higher than B. Initially the soldier moves at 2 m s^{-1} at A and in a direction perpendicular to the rope.

(a) Find the kinetic energy and speed of the soldier at B, stating any assumptions that you make.

(b) Find h, if the soldier comes to rest at C before swinging back.

(c) Explain why the tension does no work in this situation. [A]

15 The diagram shows part of the track of a roller coaster ride, which has been modelled as a number of straight lengths of track. The roller coaster's carriages are modelled as a particle of mass 400 kg, which can negotiate the bends A, B, C and D without any loss of speed. The speed of the roller coaster at A is 3 m s^{-1} and at B it is 10 m s^{-1}.

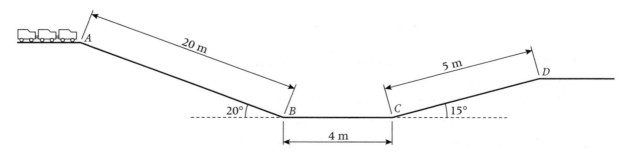

(a) Show that the work done against the resistance forces, as the roller coaster moves from A to B, is approximately 8610 J and use this to find the magnitude of the resistance forces, assuming that they are constant.

(b) Using the magnitude of the resistance forces found in (a), show that the speed of the roller coaster at D is approximately 7.4 m s^{-1} and find how far along the horizontal track beyond D it could travel before stopping.

(c) Describe **two** ways in which this model could be improved. [A]

3.4 Hooke's law

This section will consider the work done by variable forces, but will first introduce Hooke's law, which predicts the tension in a string or spring. Hooke's law will be used extensively in the later sections of this chapter.

Hooke's law provides a simple but effective model for the tension in a spring.

> Hooke's law simply states that the tension T is given by,
> $$T = \frac{\lambda x}{l},$$
> where λ is a constant called the modulus of elasticity, l is the natural or unstretched length of the spring and x is the extension of the spring. The modulus of elasticity depends on the material that the spring is made from and the way in which it has been constructed.

Note that when working with Hooke's law all lengths should be in metres and that the units of λ are newtons.

If a spring is compressed instead of stretched, then Hooke's law can be used to calculate the thrust exerted by the spring.

Hooke's law can also be applied to elastic strings.

Worked example 3.10

A particle, of mass 5 kg, is suspended from a spring, of natural length 0.2 m and modulus of elasticity 40 N. Find the extension of the spring when the particle is in equilibrium.

Solution

The diagram shows the forces acting on the particle. In equilibrium, the upward tension will balance the weight of the particle. This gives

$$T = 5 \times 9.8$$
$$= 49 \text{ N}$$

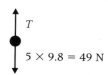

Using Hooke's law this becomes;

$$\frac{40x}{0.2} = 49$$
$$x = 0.245 \text{ m}$$

Worked example 3.11

A spring has natural length 0.5 m. A 2 kg mass is suspended from the spring and in equilibrium the extension of the spring is 0.05 m. Find the modulus of elasticity of the spring.

Solution

The diagram shows the forces acting on the mass. When it is in equilibrium, we have;

$$T = 2 \times 9.8 = 19.6 \text{ N}$$

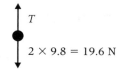

Using Hooke's law this becomes;

$$\frac{\lambda \times 0.05}{0.5} = 19.6$$
$$\lambda = 196 \text{ N}$$

EXERCISE 3D

1. A spring has modulus of elasticity 40 N and natural length 0.8 m. A particle is attached to the end of the spring and the system is allowed to hang vertically. Find the extension of the spring when the particle is in equilibrium, if the mass of the particle is:
   (a) 2 kg
   (b) 1.2 kg
   (c) 200 grams.

2. A spring has natural length 0.25 m and modulus of elasticity 20 N. A force of magnitude 4 N is applied to one end, while the other remains fixed. Find the extension of the spring when the forces are in equilibrium.

3. A spring has natural length 20 cm. When it supports a particle of mass 4 kg in equilibrium, it has an extension of 5 cm. Find the modulus of elasticity of the spring.

4. An elastic string has natural length 60 cm and modulus of elasticity 4 N. It stretches 10 cm when it supports an object with an unknown mass in equilibrium. Find the mass of the object.

5. Two identical springs (of negligible weight) have natural length 8 cm and modulus of elasticity 20 N. A 100 gram mass is attached to the springs so that it is in equilibrium.
   (a) Find the extension of a single spring that supports the mass.
   (b) If the springs support the mass as shown in the diagram, find the extension of each spring.
   (c) If the springs are joined end to end and then support the mass, find the total extension of springs.

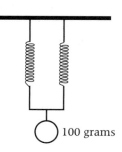

3.5 Energy and variable forces

So far our considerations have been restricted to forces that either remain constant or that are modelled as being constant. In this section you will consider how to extend the ideas previously encountered to variable forces.

For constant forces you have calculated the work done by the force using

Work done = Fs.

For a variable force, $F(x)$, the work done is given by

Work done = $\int F(x)\,dx$.

Comparing graphs of F against x shows how this result is derived.

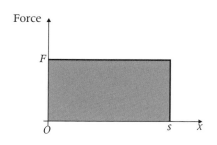

Area = Fs Area = $\int_0^s F(x)\,dx$.

Note that to use this result the force must be expressed in terms of x, the displacement of the particle.

If a variable force, $F(x)$, acts as a particle moves from $x = x_1$ to $x = x_2$, then

$$\int_{x_1}^{x_2} F(x)\,dx = \text{Work done}.$$

Using

Work done = Change in kinetic energy

gives

$$\int_{x_1}^{x_2} F(x)\,dx = \frac{1}{2}mv^2 - \frac{1}{2}mu^2.$$

Elastic potential energy

When a spring is stretched, work is done. In the same way as when work is done lifting a body it gains potential energy, a stretched or compressed spring also has potential energy, that could be converted to kinetic energy if the spring is released.

Consider a spring. The tension in the spring is given by $T = \dfrac{\lambda x}{l}$.

If the spring is extended by a distance e from its natural length, then the work done will be given by

$$\int_0^e \frac{\lambda x}{l} dx = \left[\frac{\lambda x^2}{2l}\right]_0^e$$

$$= \frac{\lambda e^2}{2l}$$

As this is the work done in stretching the spring, you can state that this is also the amount of potential energy stored in the spring. This is often expressed as follows.

> The elastic potential energy (EPE) of a stretched (or compressed) spring $= \dfrac{\lambda e^2}{2l}$.

The following examples show how this result can be applied.

Worked example 3.12

A spring has natural length 20 cm and modulus of elasticity 80 N. Calculate the work done in stretching the spring:

(a) from its natural length to a length of 25 cm,

(b) from a length of 30 cm to 40 cm.

Solution

(a) This is found by substituting $\lambda = 80$, $l = 0.2$ and $e = 0.05$ into the formula $\dfrac{\lambda e^2}{2l}$

$$\text{Work done} = \frac{80 \times 0.05^2}{2 \times 0.2}$$

$$= 0.5 \text{ J}$$

(b) The required amount is the work done to stretch the spring to 40 cm less the work done to stretch it to 30 cm. Note that the extensions will be 0.1 m and 0.2 m.

$$\text{Work done} = \frac{80 \times 0.2^2}{2 \times 0.2} - \frac{80 \times 0.1^2}{2 \times 0.2}$$

$$= 8 - 2$$

$$= 6 \text{ J}$$

Worked example 3.13

A ball, of mass 300 grams, is placed on top of a spring of natural length 10 cm and modulus of elasticity 80 N. The spring is compressed until its length is 5 cm and released. Find the maximum height of the ball above the base of the spring.

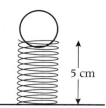

Solution

The initial energies can be calculated.

The EPE of the spring $\dfrac{80 \times 0.05^2}{2 \times 0.1} = 1$ J

The gravitational potential energy of the ball is
$0.3 \times 9.8 \times 0.05 = 0.147$ J.

So the total initial energy of the ball is 1.147 J.

At its highest point the gravitational potential energy is
$0.3 \times 9.8 \times h = 2.94h$.

If you assume that energy is conserved, then

$$2.94h = 1.147$$
$$h = \dfrac{1.147}{2.94} = 0.390 \text{ m (to 3 sf)}$$

Worked example 3.14

A sphere, of mass 200 grams, is attached to one end of an elastic string. The other end of the string is fixed to the point O. The string has natural length 50 cm and modulus of elasticity 4.9 N. The sphere is released from rest at O and falls vertically.

(a) Calculate the maximum distance between the sphere and O.

(b) Determine the maximum speed of the sphere.

Solution

(a) Let e be equal to the extension of the string, so the sphere falls $0.5 + e$ m. Then the total gravitational potential energy lost as the sphere falls is given by

$$0.2 \times 9.8 \times (0.5 + e) = 0.98 + 1.96e$$

At the sphere's lowest point, the elastic potential energy is

$$\text{EPE} = \dfrac{4.9e^2}{2 \times 0.5}$$
$$= 4.9e^2$$

As energy is conserved the final EPE will be equal to the gravitational potential energy lost, so

$$0.98 + 1.96e = 4.9e^2$$
$$0 = 4.9e^2 - 1.96e - 0.98$$

Solving this quadratic equation gives two values of e

$$e = 0.690 \text{ m or } e = -0.290 \text{ m (to 3 sf)}$$

As the second of these does not apply, because $e \geq 0$, there is a maximum extension of 0.690 m.

The maximum distance between the sphere and the point of suspension is then

$$0.690 + 0.5 = 1.190 \text{ m}$$

(b) The sphere will reach its maximum speed when its acceleration becomes zero. This will happen when the sphere reaches its equilibrium position. After this the sphere will decelerate.

First find the equilibrium position. If e is the extension of the spring at the equilibrium position, then

$$0.2 \times 9.8 = \frac{4.9e}{0.5}$$

$$e = \frac{0.98}{4.9} = 0.2 \text{ m}$$

Using this value for the extension of the string and v for the speed of the sphere

$$\text{EPE} = \frac{4.9 \times 0.2^2}{2 \times 0.5}$$

$$= 0.196 \text{ J}$$

$$\text{KE} = \tfrac{1}{2} \times 0.2v^2$$

$$= 0.1v^2$$

$$\text{GPE lost} = 0.2 \times 9.8 \times (0.5 + 0.2)$$

$$= 1.372 \text{ J}$$

As the total energy will remain equal to the initial energy of the system

$$\text{GPE lost} = \text{EPE gained} + \text{KE gained}$$

$$1.372 = 0.196 + 0.1v^2$$

$$1.176 = 0.1v^2$$

Then solving for v gives

$$v = \sqrt{11.76}$$

$$= 3.43 \text{ m s}^{-1} \text{ (to 3 sf)}$$

EXERCISE 3E

1 An elastic spring has natural length 2.5 m and modulus of elasticity 100 N. Calculate the work done in extending it:

  (a) from 2.5 m to 2.7 m,

  (b) from 2.7 m to 2.9 m.

2. In a horizontal pinball machine the spring, which has natural length 20 cm, is compressed 5 cm. If the mass of the ball is 20 grams and the modulus of elasticity of the spring is 80 N. What is the speed of the ball when it leaves the spring assuming that friction can be neglected?

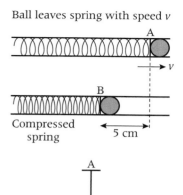

Ball leaves spring with speed v

3. A 100 gram mass is attached to the end B of an elastic string AB with modulus of elasticity of 3.92 N and natural length 0.25 m, the end A being fixed. The mass is pulled down from A until AB is 0.5 m and then released.

   Find the velocity of the mass when the string first becomes slack and show that the mass comes to rest when it reaches A.

4. A 100 grams mass is attached to the end of an elastic string which hangs vertically with the other end fixed.

   The string has a modulus of elasticity of 1.96 N and natural length 0.25 m. If the mass is pulled downwards until the length of the string is 0.5 m and released, show that the mass comes to rest when the string becomes slack.

5. A block, of mass 4 kg, is attached to one end of a length of elastic string. The other end of the string is fixed to a wall. The block is placed on a horizontal surface as shown in the diagram.

   The elastic string has natural length 60 cm and modulus of elasticity 60 N. The block is pulled so that it is 1 m from the wall and is then released from rest.

   (a) Calculate the elastic potential energy when the block is 1 m from the wall.

   (b) If the surface is smooth, show that the speed of the block when it hits the wall is 2 m s^{-1}.

   (c) The surface is in fact rough and the coefficient of friction between the block and the surface is 0.3.
   (i) Show that the speed of the block when the string becomes slack is approximately 1.28 m s^{-1}.
   (ii) Determine whether or not the block will hit the wall. [A]

6. A mass m is attached to one end B of an elastic string AB of natural length l. The end A of the string is fixed and the mass falls vertically from rest at A. In the subsequent motion, the greatest depth of the mass below A is $3l$. Calculate the modulus of elasticity of the string.

7 An energy-absorbing car bumper is modelled as a spring of natural length 20 cm and modulus of elasticity 105 kN. The 1200 kg car approaches a massive wall with a speed of 8 kph. Modelling the car as the mass spring system shown in the diagram, determine:

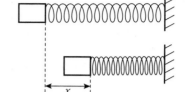

(a) the velocity of the car during contact with the wall when the spring is compressed a distance x m,

(b) the maximum compression of the spring.

8 An elastic string, AB, has natural length 1 m and its modulus of elasticity is 25 N. The end A is fixed to a smooth horizontal surface and a particle of mass 4 kg is attached to the end B. The string is stretched so that the particle is 1.5 m from A and on the surface. The particle is released from rest.

Show that the speed of the particle when the string becomes slack is 1.25 m s^{-1}. [A]

9 An elastic string has modulus of elasticity 12 N and natural length 0.5 m. A particle of mass 0.5 kg is attached to one end of the string. The other end of the string is attached to a fixed point P. The particle is pulled down until it is 1.5 m below P.

(a) Calculate the elastic potential energy of the string when the particle is 1.5 m below P.

(b) The particle is released.

   (i) Show that the kinetic energy of the particle is 7.1 J, when the string becomes slack.

   (ii) Find the kinetic energy of the particle when it is 0.5 m above P.

   (iii) Find the maximum height of the particle above P. [A]

10 A child's toy consists of a sphere mounted on top of a light spring, inside a smooth tube. The sphere is pushed down, so that the spring is compressed. When released, the sphere moves vertically upwards. The toy is shown in the diagram.

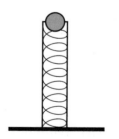

Model the sphere as a particle of mass 30 grams. The spring has a natural length of 3 cm and a modulus of elasticity of 4.5 N, and rests on a horizontal surface.

(a) Find the compression of the spring when the sphere is at rest in equilibrium.

The spring is now compressed until its length is 1 cm. The system is then released.

(b) Find the maximum height of the sphere in the subsequent motion. [A]

11 A bungee jumper, of mass 70 kg, is attached to one end of a light elastic cord of natural length 14 m and modulus of elasticity 2744 N. The other end of the cord is attached to a bridge, approximately 40 m above a river.

The bungee jumper steps off the bridge at the point where the cord is attached and falls vertically. The bungee jumper can be modelled as a particle throughout the motion. Hooke's law can be assumed to apply throughout the motion.
   (a) Find the speed of the bungee jumper at the instant the cord first becomes taut.
   (b) The cord extends by x m beyond its natural length before the bungee jumper first comes instantaneously to rest.
       (i) Show that $x^2 - 7x - 98 = 0$.
       (ii) Hence find the value of x.
       (iii) Calculate the deceleration experienced by the bungee jumper at this point. [A]

12 A bungee jumper, of mass 80 kg, is attached to an elastic rope of natural length 20 m and modulus of elasticity 2000 N. The other end of the elastic rope is attached to a bridge. The bungee jumper steps off the bridge at the point where the rope is attached and falls vertically. When the bungee jumper has fallen x m, his speed is v m s^{-1}.
   (a) By considering energy, show that when x exceeds a certain minimum value $20v^2 = 1392x - 25x^2 - 10\,000$ and state this minimum value of x.
   (b) Find the maximum value of x.
   (c) (i) Show that the speed of the bungee jumper is a maximum when $x = 27.84$ m.
       (ii) Hence find the maximum speed of the bungee jumper. [A]

13 A 'reverse bungee jump' consists of a 12 m length of elastic rope, that is stretched into a V-shape, ABC, on a frame, as shown in the diagram. The ends of the elastic rope are fixed to the frame at the points A and C.

A student, of mass 85 kg, is attached to the midpoint of the elastic rope at B. The modulus of elasticity of the elastic rope is 1500 N.

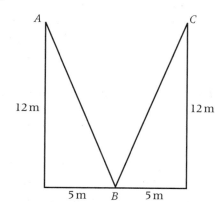

   (a) Show that the elastic potential energy of the elastic rope in the initial position shown in the diagram is 12 250 J.

The middle of the rope is then released from B and the student moves vertically upwards.
   (b) Find the speed of the student, when at a height of 12 m above B.

The student reaches his maximum height before the rope becomes taut again.
   (c) Find the maximum height of the student above B during the motion. [A]

14 A 15 tonne wagon travelling at 3.6 m s^{-1} is brought to rest by a buffer (a spring) having a natural length of 1 m and modulus of elasticity of 7.5 × 10^5 N. Assuming that the wagon comes into contact with the buffer smoothly without rebound, calculate the compression of the buffer if there is a constant rolling friction force of 800 N.

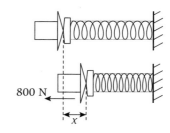

15 A 10 kg block rests on a rough horizontal table. The spring, which is not attached to the block, has a natural length of 0.8 m and modulus of elasticity 400 N. If the spring is compressed 0.2 m and then released from rest determine the velocity of the block when it has moved through 0.4 m. The coefficient of friction between the block and the table is 0.2.

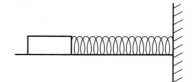

16 A platform P has negligible mass and is tied down so that the 0.4 m long cords keep the spring compressed 0.6 m when nothing is on the platform. The modulus of elasticity of the spring is 200 N. If a 2 kg block is placed on the platform and released when the platform is pushed down 0.1 m.
   (a) Show that the spring loses 13 J of elastic potential energy by the time the cords become taut.
   (b) Find the speed of the block when it leaves the platform.
   (c) Find the maximum height reached by the block.

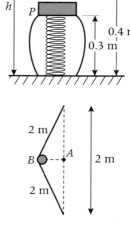

17 An archaeologist investigates the mechanics of large catapults used in sieges of castles. The diagram shows a simplified plan of such a catapult about to be fired horizontally.

The rock B of mass 20 kg is in the catapult as shown. Calculate the speed with which the rock is released at A when the elastic string returns to its natural length of 2 m if the string's modulus of elasticity is 5000 N.

18 An elastic string has natural length 2 m and modulus of elasticity λ N. One end of the string is fixed at the point O, and a particle of mass 20 kg is attached to the other end of the string.
   (a) When in equilibrium the particle is 2.7 m below O. Show that $\lambda = 560$ N.
   (b) The particle is now held at O and released from rest. The maximum length of the string in the subsequent motion is L.
      (i) Show that L satisfies the equation
         $5L^2 - 27L + 20 = 0$.
      (ii) Find the maximum length of the string.   [A]

3.6 Power

Power is defined as the rate of doing work. For example we may talk about a more powerful car or motorbike. This is a way of describing how quickly they gain kinetic energy. The more quickly kinetic energy is gained, the shorter the time that the work is done in and so the more powerful the vehicle.

A simple definition of power is:
$$\text{Power} = \frac{\text{work done}}{\text{time taken}}.$$

Worked example 3.15

Hannah, who has mass 50 kg, climbs a flight of stairs in 20 seconds. As she climbs the stairs she rises a total of 4 m.

(a) Calculate the work done as she climbs the stairs.
(b) Calculate her power output as she climbs the stairs.

Solution

(a) Work done $= 50 \times 9.8 \times 4$
$\phantom{\text{Work done}} = 1960 \text{ J}$

(b) Power $= \dfrac{\text{work done}}{\text{time taken}}$

$\phantom{\text{Power}} = \dfrac{1960}{20}$

$\phantom{\text{Power}} = 98 \text{ W}$

Note that the SI units for power are watts (W). An alternative unit is J s^{-1}. Another, more traditional, unit that is sometimes used for power is the horsepower (hp), and is such that 1 hp is approximately 740 watts.

Work done by force

An alternative approach to power is to derive a formula based on the definition that power is the rate of doing work.

The work done by a force is Fs, where F is the magnitude of the force and s is the displacement. As rates can found by differentiating with respect to t, we have

$$\text{Power} = \frac{d}{dt}(Fs).$$

Considering the case of a constant force leads to

$$\text{Power} = \frac{d}{dt}(Fs)$$

$$= F\frac{ds}{dt}$$

$$= Fv$$

This result is very useful and can be applied in many examples.

Worked example 3.16

A car experiences a resistance force of magnitude 1200 N, when travelling at a constant speed of 25 m s^{-1}. Calculate the power output of the car.

Solution

As the car is travelling at a constant speed, there must be a forward force on the car equal in magnitude to the resistance force. So in this case the force F exerted by the car has magnitude 1200 N and $v = 25$ m s^{-1}. Using $P = Fv$, gives

$$P = 1200 \times 25$$
$$= 30\,000 \text{ W}$$

Worked example 3.17

A car, of mass 1200 kg, experiences a resistance force that is proportional to its speed. The car has a maximum power output of 36 000 W and a maximum speed of 40 m s^{-1}.

(a) Determine an expression for the magnitude of the resistance force, when the speed of the car is v m s^{-1}.

(b) Find the power output if the car is accelerating at 2 m s^{-2} and is travelling at 10 m s^{-1}.

(c) Calculate the maximum acceleration of the car when it is travelling at 20 m s^{-1}.

Solution

(a) As the resistance is proportional to the speed we have
$$R = kv.$$
At its top speed the resistance force will be equal in magnitude to the forward force exerted by the car, so $F = kv$. Using $P = Fv$ gives

$$36\,000 = 40k \times 40$$
$$k = \frac{36\,000}{40^2} = 22.5$$

So the resistance force is $R = 22.5v$ N.

(b) The car exerts a forward force, of magnitude F N, and experiences a resistive force of magnitude $22.5 \times 10 = 225$ N. So the resultant force on the car is $F - 225$. As the car is accelerating at 2 m s^{-2}, you can apply Newton's second law to give

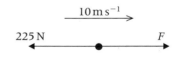

$$F - 225 = 1200 \times 2$$
$$F = 2625 \text{ N}$$

Now the power output can be found using $P = Fv$ as
$$P = 2625 \times 10$$
$$= 26\,250 \text{ W}$$

(c) At a speed of 20 m s^{-1}, the car will experience a resistance force of 450 N. If it exerts a forward force of magnitude F N and accelerates at a m s^{-2}, then Newton's second law can be applied to give

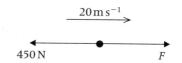

$$F - 450 = 1200a$$
$$F = 1200a + 450$$

Now using the relationship $P = Fv$, you can form and solve an expression for a.

$$36\,000 = (1200a + 450) \times 20$$
$$1800 = 1200a + 450$$
$$a = \frac{1800 - 450}{1200}$$
$$= 1.125 \text{ m s}^{-2}$$

Worked example 3.18

A car has a maximum power of 32 000 W and a mass of 1200 kg. The resistance forces acting on the car are modelled as being proportional to the speed of the car. The car has a top speed of 40 m s^{-1}.

(a) Find the constant of proportionality in the model for the resistance forces.

(b) Find the maximum speed of the car up a slope that is inclined at an angle α to the horizontal, where $\sin \alpha = \frac{1}{14}$.

(c) Find the power output of the car if it is accelerating at 2 m s^{-2}, while travelling at 5 m s^{-1}, on a horizontal road.

Solution

(a) At top speed the resistance forces are equal to the forward forces F. Using $P = Fv$, and assuming that the magnitude of the resistance force is kv, gives

$$32\,000 = k \times 40^2$$
$$k = \frac{32\,000}{40^2}$$
$$= 20$$

(b) When travelling up the hill at a constant speed, the resultant force must balance both the component of gravity down the slope and the resistance force. So the force, F, exerted by the car is given by

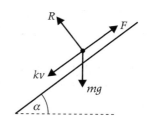

$$F = mg \sin \alpha + kv$$
$$= 1200 \times 9.8 \times \tfrac{1}{14} + 20v$$
$$= 840 + 20v$$

Using $P = Fv$ gives

$$32\,000 = (840 + 20v)v$$
$$20v^2 + 840v - 32\,000 = 0$$
$$v^2 + 42v - 1600 = 0$$

Solving this quadratic gives:

$$v = \frac{-42 \pm \sqrt{42^2 - 4 \times 1 \times (-1600)}}{2 \times 1}$$
$$= 24.2 \text{ or } -66.2 \text{ m s}^{-1}$$

So the maximum speed up the slope is 24.2 m s^{-1}.

(c) If the car is travelling at 5 m s^{-1}, then there will be a resistance force of 100 N:

$$F - 100 = 1200 \times 2$$
$$= 2500 \text{ N}$$

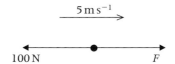

The power can then be found using $P = Fv$:

$$P = 2500 \times 5$$
$$= 12\,500 \text{ W}$$

EXERCISE 3F

1. A crane lifts a load, of mass 800 kg, to a height of 12 m in 2 minutes.
   (a) Calculate the work done by the crane.
   (b) Find the power of the crane.

2. A child, of mass 56 kg, climbs up a flight of stairs in 49 seconds. There are 50 steps, each of height 18 cm. Calculate the rate at which the child was working as she climbed the stairs.

3. A train travels at a constant speed of 30 m s^{-1} and experiences a resistance force of magnitude 30 000 N at this speed. Calculate the power output of the train.

4. A pump is used to raise water from ground level into a tank at a height of 5m. The pump is able to pump 5000 litres per hour. Find the power of the pump. (The mass of 1 litre of water is 1 kg.)

5. A car, of mass 1000 kg, has a maximum power output of 36 000 W and a maximum speed of 40 m s^{-1}. The resistance force on the car is proportional to its speed.
   (a) Find the magnitude of the resistance force when it is travelling at its maximum speed.
   (b) Find the magnitude of the resistance force when the car is travelling at 30 m s^{-1}.
   (c) Find the maximum possible acceleration of the car when it is travelling at 30 m s^{-1}.

6 A lorry of mass 5000 kg moves in a straight line along a horizontal road. During this motion the power of the lorry's engine is 40 kW.

When the lorry is travelling at a speed of 10 m s^{-1}, the resistance to the motion of the lorry is of magnitude 2000 N. At this speed find:

(a) the tractive force produced by the engine,

(b) the acceleration of the lorry. [A]

7 A motorcycle, of mass 300 kg, has a maximum power output of 30 kW and a top speed of 40 m s^{-1} on the horizontal. Assume that the resistance forces that act on the motorcycle are proportional to its speed.

(a) Find an expression for the magnitude of the resistance force acting on the motorcycle when it is travelling at v m s^{-1}.

(b) Calculate the power output of the motorcycle if it travels at a constant 25 m s^{-1}.

(c) Find the maximum possible speed of the motorcycle, while it is accelerating at 2 m s^{-2}.

8 A car, of mass 1000 kg, is assumed to experience a resistance force that is proportional to its speed. The car has a maximum power output of 50 000 W and a top speed of 50 m s^{-1} on the horizontal. A slope is inclined at an angle α to the horizontal, where $\sin \alpha = \frac{1}{10}$. Find the maximum speed of the car when travelling from rest up or down the slope.

9 A motorcycle, of mass 300 kg, has a maximum power output of 30 kW and a top speed of 45 m s^{-1} on the horizontal. Assume that the resistance forces that act on the motorcycle are proportional to its speed.

(a) Find an expression for the magnitude of the resistance force that acts on the motorcycle at a speed of v m s^{-1}.

(b) Find the maximum possible speed of the motorcycle as it travels up a slope inclined at 5° to the horizontal.

(c) What would be the maximum speed of the motorcycle down the same slope?

10 A car, of mass 1200 kg, moves on a straight, horizontal road. A resistance force acts on the car and its magnitude is $40v$ N, where v is the speed of the car in metres per second.

(a) The car travels 200 m at a constant speed of 30 m s^{-1}. Find the work done in overcoming the resistance force.

(b) (i) If the car accelerates at 2 m s^{-2} while moving at 20 m s^{-1}, show that the power of the car is 64 000 W.

(ii) Given that the maximum power of the the car is 64 000 W, find the maximum speed of the car. [A]

11 A car of mass 1200 kg is being driven up a straight road inclined at 5° to the horizontal. Resistive forces acting on the car total 1960 N.

(a) Draw a diagram showing all the forces acting on the car.

(b) The car is moving with constant speed 15 m s^{-1}.
   (i) Show that the tractive force produced by the engine is approximately 2985 N.
   (ii) Determine the rate at which the engine is doing work.

(c) The engine has a maximum power output of 60 kW. Assuming resistive forces are constant, find the maximum possible speed of the car up the same slope, if it is initially at rest. [A]

12 A car of mass 1200 kg, experiences a resistance force of magnitude $40v$ N when travelling at v m s^{-1}.
The car travels up a slope inclined at an angle $\sin^{-1}\left(\frac{1}{10}\right)$ to the horizontal. When its speed is 20 m s^{-1} the car is accelerating at 1 m s^{-2}.

(a) Show that the power output of the car is 63 520 W.

(b) Assume the power calculated in (a) is the maximum for the car. The driver of the car finds that, when travelling up a different slope, the maximum speed of the car is 25 m s^{-1}. Find the angle between this slope and the horizontal. [A]

13 A car, of mass 1000 kg, has a maximum speed of 40 m s^{-1} on a straight horizontal road. When the car travels at a speed v m s^{-1}, it experiences a resistance force of magnitude $35v$ N.

(a) Show that the maximum power of the car is 56 000 W.

(b) The car is travelling on a straight horizontal road. Find the maximum possible acceleration of the car when its speed is 20 m s^{-1}.

(c) The car starts from rest on a slope inclined at 5° to the horizontal. Find the maximum possible speed of the car as it travels in a straight line up this slope. [A]

14 A car, of mass 1000 kg, has a power output of 24 000 W. It experiences a resistance force of magnitude $20v$ when travelling at a speed v.

(a) Find the maximum speed of the car on a horizontal road.

(b) Find the maximum speed of the car when travelling up a slope inclined at an angle α to the horizontal, where $\sin \alpha = \frac{1}{20}$.

15 A car, of mass 1000 kg, is assumed to experience a resistance force that is proportional to its speed squared. The car has a maximum power output of 32 000 W and a top speed of 40 m s^{-1}, on the horizontal.
  (a) Find the resistance force acting when the car travels at a speed of 20 m s^{-1}.
  (b) The car travels 500 m up a slope inclined at an angle α to the horizontal, where $\sin \alpha = \frac{1}{15}$. The car travels at a constant speed of 20 m s^{-1}. Find the work done by the car as it travels up the slope.

16 A cyclist can pedal up a slope, inclined at 4° to the horizontal, at a maximum speed of 2 m s^{-1}. Model the cycle as a particle of mass 70 kg. Assume that there are no resistance forces acting on the cyclist.
  (a) When he is pedalling up the slope at his maximum speed, show that the power output of the cyclist is approximately 96 W.
  (b) If the power output remains the same, find the maximum speed of the cyclist when travelling up a slope inclined at 6° to the horizontal.
  (c) (i) When modelling the motion uphill, explain why it is reasonable to assume that there is no resistance.
      (ii) When trying to find the maximum speed of the cyclist down the slope, explain why it is **not** reasonable to assume that there is no resistance. [A]

17 A car, of mass 1200 kg, has a maximum power output of 48 000 W. On a horizontal road the car has a maximum speed of 40 m s^{-1}. Assume that the resistance forces acting on the car are proportional to its speed.
  (a) Find the resistance force acting on the car when it travels at v m s^{-1}.
  (b) Find the percentage reduction in the power output of the car if its speed is reduced by 10%.
  (c) Use your answer to **(b)** to describe one advantage of reducing the speed at which the car is driven.
  (d) Find the maximum speed of the car, when being driven up a slope at 4° to the horizontal. [A]

18 The maximum power output of a car is 50 000 W and its top speed, on a horizontal road, is 40 m s^{-1}. In order to model the motion of the car, assume that it experiences a resistance force proportional to its speed.
  (a) Find the resistance force when the car is travelling at 20 m s^{-1}.
  (b) When the car tows a caravan the resistance force is increased by 50%. Find the maximum speed of the car when it tows the caravan on a horizontal road. [A]

19 A car and its driver, of a total mass 500 kg, are ascending a hill of inclination $\sin^{-1}(\frac{1}{7})$ to the horizontal with a constant speed of 5 m s^{-1}. Given that the motion is opposed by a frictional force of magnitude 800 N, find the power generated by the engine of the car.

The driver presses the accelerator, which has the effect of suddenly increasing the power to 20 kW. Calculate the resulting acceleration of the car. [A]

20 A car, of mass 1000 kg, travels up a hill inclined at 4° to the horizontal. Assume that the car experiences a constant resistance force, and is moving at constant or increasing speed.

(a) Draw and label a diagram to show the forces acting on the car, if it is modelled as a particle. Describe one weakness of modelling the car as a particle.

(b) If the car exerts a forward force of 1000 N when travelling at a constant speed up the hill, show that the magnitude of the resistance force is 316 N to the nearest newton.

(c) Use your answer to (b) to find the power output of the car if it is travelling at 20 m s^{-1} and accelerating at 1.5 m s^{-2}. [A]

Key point summary

1. The kinetic energy of a body is defined as $\frac{1}{2}mv^2$.   *p32*

2. The work done by the force is the quantity Fs.   *p34*

3. Gravitational potential energy $= mgh$.   *p40*

4. Hooke's law states that the tension T is given by $\frac{\lambda x}{l}$.   *p47*

5. The elastic potential energy (EPE) of a stretched or compressed spring $= \frac{\lambda e^2}{2l}$.   *p50*

6. Power $= \dfrac{\text{work done}}{\text{time taken}} = $ Rate of doing work   *p57*

Formulae to learn

Kinetic energy	$\frac{1}{2}mv^2$
Work done	Fs or $Fs\cos\theta$
Work done = change in kinetic energy	
Gravitational potential energy	mgh
Hooke's law	$T = \dfrac{\lambda x}{l}$
Work done by a variable force	$\int F(x)\,dx$
Elastic potential energy	$\dfrac{\lambda e^2}{2l}$
Power	$P = Fv$

Test yourself	**What to review**
1 Calculate the gain in kinetic energy as a car, of mass 1200 kg, increases in speed from 10 to 25 m s^{-1}.	Section 3.1
2 A force of 500 N acts on a car, of mass 1250 kg, in the direction of motion. The car has an initial speed of 8 m s^{-1}. The force acts as the car travels 25 m on a horizontal surface. (a) Calculate the work done by the force as the car travels this distance. (b) Find the final speed of the car.	Section 3.2
3 A go-kart, of mass 50 kg, is at rest at the top of a slope inclined at 8° to the horizontal. Find the speed of the go-kart when it has travelled 70 m down the slope: (a) if there is no resistance to the motion, (b) there is a constant 20 N resistance force.	Section 3.3
4 A 'dropslide' at a leisure park consists of a curved section AB and a horizontal section BC shown below. Children start at rest at the point A and slide down to B and on towards C. The points, A, B and C all lie in the same vertical plane and the motion of the child is in this plane. A child of mass 30 kg uses the slide. (a) Assuming that there is no friction or air resistance acting on the child on the section AB, find the speed of the child at B. How would this speed compare with that of a heavier child? (b) If the child travels 5 m along BC before stopping, find the magnitude of the friction force between the child and the surface. State two factors that would influence the magnitude of this force for different children. [A]	Section 3.3

Test yourself (continued)

What to review

5 A sphere, of mass 500 grams, is attached to an elastic string of natural length 70 cm and modulus of elasticity 70 N. One end of the string is fixed to a point O.

(a) Find the length of the elastic when the sphere hangs in equilibrium.

The sphere is released from the point O.

(b) Find the maximum distance between the sphere and O.

(c) Find the speed of the sphere when the length of the elastic string is 80 cm.

(d) Find the maximum speed of the sphere.

Section 3.5

6 A car of maximum power output 32 kW and mass 800 kg, is travelling up a slope at an angle θ to the horizontal.

A simple model for the motion of the car assumes that there are no resistance forces acting on the car.

(a) Find the angle θ, to the nearest degree, for the steepest slope that the car can ascend at a speed of 10 m s^{-1}.

A refined model for the motion of the car would take account of the resistance forces on the car.

(b) The slope is in fact at 5° to the horizontal, and the car is still travelling at its maximum speed of 10 m s^{-1}. Find the magnitude of the resistance forces on the car.

(c) The resistance forces on the car are assumed to be proportional to the speed. Use your result to **(b)** to find a simple model for the resistance forces. [A]

Section 3.6

Test yourself ANSWERS

1 315 000 J.

2 (a) 12 500 J; (b) 9.17 m s^{-1}.

3 (a) 13.8 m s^{-1}; (b) 11.6 m s^{-1}.

4 (a) 7.92 m s^{-1}, same; (b) 188 N, mass, clothes.

5 (a) 74.9 cm; (b) 1.015 m; (c) 3.70 m s^{-1}; (d) 3.77 m s^{-1}.

6 (a) 24°; (b) 2517 N; (c) 252 N.

Kinematics and variable acceleration

Learning objectives

After studying this chapter you should be able to:
- differentiate displacements or position vectors to give velocities and accelerations for one, two or three dimensions
- integrate accelerations to give velocities and position vectors or displacements.

4.1 Introduction

In the M1 module you would have considered only cases where the acceleration of an object is constant. For example a ball falling under gravity or perhaps a car with a constant acceleration. However there are many situations where it is unrealistic to assume that the acceleration of a body is constant or where better solutions to problems can be obtained by modelling the acceleration as variable. For example circular motion involves an acceleration that is always changing direction and the acceleration of a car may decrease as it gains speed.

In some cases the acceleration can be expressed as a function of time, in others it may depend on the speed or velocity of the object under consideration. In this chapter you will consider cases where the acceleration is dependent on time.

This chapter will require you to use calculus instead of the constant acceleration equations. From now on when approaching a problem it is important to decide whether or not the acceleration is constant before using the constant acceleration equations. In general you will find that there will be relatively little use of constant acceleration equations in this module.

4.2 Displacement to velocity and acceleration

In the M1 module it was noted that the velocity was given by the gradient of a displacement–time graph and that the acceleration was given by the gradient of a velocity–time graph. These results can be used as the basis of your work with variable acceleration.

The gradient of a curve is given by its derivative so we can deduce that the velocity is given by the derivative, with respect to time, of the displacement.

> $v = \dfrac{dx}{dt}$
>
> The velocity is equal to the rate of change of the displacement.

Similarly the acceleration will be given by the derivative of the velocity, with respect to time.

> $a = \dfrac{dv}{dt}$
>
> The acceleration is equal to the rate of change of the velocity.

You can also write

$a = \dfrac{d^2x}{dt^2}.$

The following worked examples illustrate how these results can be applied.

Worked example 4.1

The height of a bullet, h metres, fired vertically upwards, at time t seconds, is given by:

$h = 3 + 80t - 4.9t^2.$

(a) Show that the acceleration of the bullet is constant.
(b) Find the maximum height reached by the bullet.

Solution

(a) First differentiate h to find the velocity.

$v = \dfrac{dh}{dt}$

$ = 80 - 2 \times 4.9t$

$ = 80 - 9.8t$

Now differentiate again to find the acceleration.
$$a = \frac{dv}{dt}$$
$$= -9.8$$

So the acceleration is constant and has magnitude 9.8 m s^{-2}.

(b) The maximum height will be attained when the velocity is zero.
$$80 - 9.8t = 0$$
$$t = \frac{80}{9.8}$$

This can then be substituted into the expression for h to find the maximum height.
$$h = 3 + 80 \times \frac{80}{9.8} - 4.9 \times \left(\frac{80}{9.8}\right)^2$$
$$= 330 \text{ m (to 3 sf)}$$

This example shows that calculus can also be applied to cases involving constant acceleration.

Worked example 4.2

As a car slows down the distance, s metres, it has travelled at time t seconds is modelled by the equation:
$$s = \frac{t^4}{8} - t^3 + 16t + 32 \text{ for } 0 \leq t \leq 4.$$

(a) Show that when $t = 4$ the car has zero velocity.

(b) Find the acceleration, when $t = 0$, $t = 2$ and $t = 4$.

(c) Describe how the acceleration of the car changes.

Solution

(a) First you need to differentiate s to find an expression for the velocity.
$$v = \frac{ds}{dt}$$
$$= \frac{4t^3}{8} - 3t^2 + 16$$
$$= \frac{t^3}{2} - 3t^2 + 16$$

Now substitute $t = 4$ in this expression.
$$v = \frac{4^3}{2} - 3 \times 4^2 + 16$$
$$= 32 - 48 + 16$$
$$= 0$$

(b) Now differentiate again to find the acceleration.

$$a = \frac{dv}{dt}$$
$$= \frac{3t^2}{2} - 6t$$

Substituting the values $t = 0$, $t = 2$ and $t = 4$ gives

$$t = 0, a = \frac{3 \times 0^2}{2} - 6 \times 0 = 0 \text{ m s}^{-2}$$

$$t = 2, a = \frac{3 \times 2^2}{2} - 6 \times 2 = -6 \text{ m s}^{-2}$$

$$t = 4, a = \frac{3 \times 4^2}{2} - 6 \times 4 = 0 \text{ m s}^{-2}$$

(c) The acceleration is a quadratic function of t with zeros when $t = 0$ and $t = 4$. As the coefficient of t^2 is positive the graph has the shape shown in the diagram. The magnitude of the acceleration increases from 0 m s^{-2} to 6 m s^{-2} and then decreases back to 0 m s^{-2} as the car comes to rest after 4 seconds.

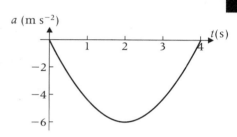

Worked example 4.3

A ball is released from rest at the top of a tall building and falls vertically. The distance fallen by the ball at time t seconds is x m where

$$x = 5t + 2.5e^{-2t} - 2.5$$

(a) Find an expression for the velocity and sketch a graph to show how the velocity varies with time.

(b) Find an expression for the acceleration and sketch a graph to show how this varies with time.

Solution

(a) Differentiate x with respect to t to find v.

$$v = \frac{dx}{dt}$$
$$= 5 - 5e^{-2t}$$

Initially v is 0, but increases towards 5 m s^{-1} as t increases, as shown in the graph.

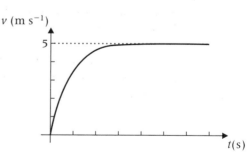

(b) Differentiating again gives the acceleration.

$$a = \frac{dv}{dt}$$
$$= 10e^{-2t}$$

Initially the acceleration is 10 m s^{-2}, but this decreases towards 0 as shown in the graph.

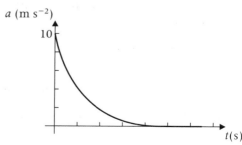

EXERCISE 4A

1 The distances, in metres, travelled by a cyclist after t seconds are given by

$$s = \frac{t^3}{6} - \frac{t^4}{120} \quad \text{for } 0 \leq t \leq 10.$$

  (a) How far has the cyclist travelled when $t = 10$.

  (b) Find an expression for the velocity of the cyclist at time t.

  (c) Find an expression for the acceleration of the cyclist at time t.

  (d) Describe how the acceleration of the cyclist changes.

2 A car accelerates from rest so that the distance that it has travelled in t seconds is s metres where $s = t^2 - \frac{t^3}{60}$.

  (a) Find expressions for the velocity and acceleration at time t seconds.

  The expression for s is valid while the acceleration is greater than or equal to zero.

  (b) Find the time when the acceleration becomes zero.

  (c) Find the velocity of the car when the acceleration is zero.

  (d) How far does the car travel before the acceleration becomes zero?

3 A lift rises from ground level. The height, s metres, of the lift at time t seconds is given by $s = \frac{3t^2}{10} - \frac{t^3}{50}$ for $0 \leq t \leq 10$.

  (a) Show that the lift comes to rest when $t = 10$.

  (b) Sketch a graph to show how the acceleration of the lift varies with time.

4 A car driver sees a red traffic light in front of him and starts to brake. The distance, s m, travelled while the car has been braking for t seconds is given by

$$s = \frac{45t}{2} - \frac{3t^2}{2} + \frac{t^3}{30}$$

This expression only applies from $t = 0$ to the time when the car comes to rest.

  (a) Find the range of values of t for which the expression for s is valid.

  (b) Find the distance travelled while the car comes to rest.

  (c) Sketch an acceleration–time graph for the car.

5 A firework manufacturer is designing a new type of firework. They want it to rise so that the height, h metres, at time t seconds is given by $h = 9t^2 - \dfrac{t^4}{12}$. The firework should explode when it reaches its maximum height.

Find the maximum height of the firework.

6 A particle moves along a straight line. At time t the displacement of the particle from its initial position is x, where
$$x = 4t + 2e^{-t} - 2.$$
(a) Find the velocity of the particle at time t.
(b) Find the acceleration of the particle at time t.
(c) Describe what happens to the acceleration of the particle as t increases. [A]

7 A weight is suspended from an elastic string. It moves up and down, so that at time t seconds the distance between the weight and the fixed end of the string is x metres, where $x = 0.8 + 0.4 \sin(0.5t)$.

(a) Find the velocity of the weight at time t.
(b) What is the maximum speed of the weight?
(c) Find the acceleration of the weight when $t = 2$.
(d) Find the range of values of the acceleration.

Remember that when trigonometrical functions are integrated or differentiated, the angles used are in radians.

8 The height, h metres, of a hot air balloon at time t seconds is modelled by
$$h = 150\left(1 - \cos\left(\dfrac{t}{800}\right)\right).$$
The model is only valid while the balloon is gaining height.

(a) State the initial height of the balloon.
(b) Find the range of values of t for which the model is valid.
(c) What is the maximum height of the balloon?
(d) What is the maximum acceleration of the balloon?

9 A particle is set in motion with an initial speed of 20 m s^{-1} on a smooth horizontal surface. It slows down due to the action of air resistance, stopping after it has travelled 15 m. A possible model for the displacement, s m, at time t seconds is $s = A(1 - e^{-kt})$, where A and k are constants.

(a) State the value of A.
(b) Find k.
(c) Sketch a graph to show how the acceleration varies with time.

10 A particle, that hangs on a spring, moves so that the displacement, x metres, from its equilibrium position at time t seconds is given by $x = 4\cos 2t + 3\sin 2t$.
   (a) Find the initial displacement of the particle.
   (b) Find the initial speed of the particle.
   (c) Show that the acceleration a m s^{-2}, satisfies the relationship $a = -4x$.

11 An object falls through a fluid so that the distance fallen, in metres, at time t seconds is given by $s = 40(4e^{-\frac{t}{4}} + t - 4)$.
   (a) Find the initial and terminal speeds of the object.
   (b) Sketch a graph to show how the acceleration of the object varies with time.

12 A particle is projected vertically, so that it moves under the influence of gravity and is subject to air resistance. The height, h metres, of the particle at time t seconds is given by
$$h = \frac{1}{k}\left(\frac{g}{k} + U\right)(1 - e^{-kt}) - \frac{gt}{k}$$
where k and U are constants. The model is only valid while the particle is moving upwards.
   (a) Show that the model is valid while
$$0 \leq t \leq \frac{1}{k}\ln\left(1 + \frac{kU}{g}\right).$$
   (b) Find the initial acceleration of the particle and sketch an acceleration–time graph for the particle.

13 A rocket that is launched at a firework display is to be modelled as a particle. The height, h metres, of the rocket at t seconds after lift-off is modelled by
$$h = \frac{5t^2}{2} - \frac{t^4}{20}.$$
The rocket rises vertically from rest and this model applies until the speed of the rocket drops to zero, when all the rocket's fuel has been used. The rocket then falls back to the ground. Assume that there is no air resistance.
   (a) Find expressions for the velocity and acceleration of the rocket as it rises.
   (b) Find the range of values of t for which the above model applies and the maximum height of the rocket in this period of time.
   (c) Describe what happens to the acceleration of the rocket while it is rising and find the maximum speed of the rocket. [A]

4.3 Acceleration to velocity and displacement

The process of differentiating to move from displacement to velocity and from velocity to acceleration can be reversed using integration.

> You can integrate an acceleration to obtain a velocity and integrate a velocity to obtain a displacement.
> $$v = \int a \, dt$$
> $$s = \int v \, dt$$

When using integration in this way it is very important to remember to include the constants of integration, which will depend on the initial velocities and positions of the objects that are under consideration.

The following examples illustrate the use of integration in this context.

Worked example 4.4

A car slows down from a speed of 30 m s^{-1}. Its acceleration, a m s^{-2} at time t seconds is given by $a = -\dfrac{t}{2}$. This expression is valid while the car is moving.

(a) Find an expression for the velocity of the car at time t.

(b) Find an expression for the distance travelled by the car at time t.

(c) Find the distance that the car travels before it stops.

Solution

(a) First integrate the acceleration to obtain the velocity, v m s^{-1}.

$$v = \int -\frac{t}{2} \, dt$$
$$= -\frac{t^2}{4} + c$$

The fact that the initial speed was 30 m s^{-1} can now be used to find c. Substituting $t = 0$ and $v = 30$ gives

$$30 = -\frac{0^2}{4} + c$$
$$c = 30$$

So the velocity at time t is
$$v = 30 - \frac{t^2}{4}$$

(b) The velocity can now be integrated to obtain the displacement, s metres.
$$s = \int 30 - \frac{t^2}{4} \, dt$$
$$= 30t - \frac{t^3}{12} + C$$

If we assume that the car starts at the origin, we can substitute $t = 0$ and $s = 0$, to determine the value of C.
$$0 = 30 \times 0 - \frac{0^3}{12} + C$$
$$C = 0$$

So the displacement at time t is given by
$$s = 30t - \frac{t^3}{12}$$

(c) The first step is to find t when the car stops.
$$0 = 30 - \frac{t^2}{4}$$
$$t^2 = 120$$
$$t = \sqrt{120}$$

This value for t can now be substituted into the expression obtained for the displacement.
$$s = 30 \times \sqrt{120} - \frac{(\sqrt{120})^3}{12}$$
$$= 219 \text{ m (to 3 sf)}$$

Worked example 4.5

As a cyclist sets off from rest the acceleration, a m s^{-2}, of the cyclist, at time t seconds is given by $a = 1 - \dfrac{t}{20}$ for $0 \leq t \leq 20$.

(a) Find expressions for the velocity and displacement of the cyclist at time t.

(b) What is the speed of the cyclist after 20 seconds?

(c) How far does the cyclist travel in the 20 seconds?

Solution

(a) The acceleration should be integrated to give the velocity.

$$v = \int 1 - \frac{t}{20}\, dt$$

$$= t - \frac{t^2}{40} + c$$

As the cyclist starts at rest, substituting $t = 0$ and $v = 0$ will give c.

$$0 = 0 - \frac{0^2}{40} + c$$

$$c = 0$$

So the velocity at time t is given by $v = t - \frac{t^2}{40}$.

This can now be integrated to give the displacement.

$$s = \int t - \frac{t^2}{40}\, dt$$

$$= \frac{t^2}{2} - \frac{t^3}{120} + C$$

If we assume that the cyclist starts at the origin, then substituting $t = 0$ and $x = 0$ will give the value of C.

$$0 = \frac{0^2}{2} - \frac{0^3}{120} + C$$

$$C = 0$$

So the displacement at time t is given by $s = \frac{t^2}{2} - \frac{t^3}{120}$.

(b) We can now substitute $t = 20$ into the expression for the velocity.

$$v = 20 - \frac{20^2}{40}$$

$$= 10 \text{ m s}^{-1}$$

The cyclist reaches a speed of 10 m s^{-1}.

(c) Substituting $t = 20$ into the expression for s will give the distance travelled.

$$s = \frac{20^2}{2} - \frac{20^3}{120}$$

$$= 133\tfrac{1}{3} \text{ m}$$

Worked example 4.6

A parachutist is initially falling at a constant speed of 30 m s^{-1}, when he opens his parachute. The acceleration, a m s^{-2}, of the parachutist t seconds after the parachute has been opened is modelled as $a = -13\mathrm{e}^{-\frac{t}{2}}$.

(a) Find expressions for the velocity and the distance fallen by the parachutist at time t.

(b) Sketch a velocity–time graph for the parachutist.

(c) The parachutist must have a speed of less than 5 m s^{-1} when he lands. Find the minimum height at which he can open his parachute.

Solution

(a) First integrate the acceleration to find the velocity.

$$v = \int -13\mathrm{e}^{-\frac{t}{2}}\,\mathrm{d}t$$

$$= 26\mathrm{e}^{-\frac{t}{2}} + c$$

Using the initial velocity, you can substitute $v = 30$ and $t = 0$ to find c.

$$30 = 26\mathrm{e}^0 + c$$

$$c = 4$$

So the velocity at time t is given by $v = 26\mathrm{e}^{-\frac{t}{2}} + 4$.

This can now be integrated to give the distance that has been fallen.

$$s = \int 26\mathrm{e}^{-\frac{t}{2}} + 4\,\mathrm{d}t$$

$$= -52\mathrm{e}^{-\frac{t}{2}} + 4t + C$$

As the distance fallen will initially be zero you can substitute $t = 0$ and $s = 0$ to find C.

$$0 = -52\mathrm{e}^0 + 4 \times 0 + C$$

$$C = 52$$

The distance fallen at time t is then

$$s = 52 - 52\mathrm{e}^{-\frac{t}{2}} + 4t$$

$$= 52(1 - \mathrm{e}^{-\frac{t}{2}}) + 4t$$

(b) The expression for the velocity shows that it decreases exponentially from 30 m s^{-1} towards 4 m s^{-1}. This is shown in the graph.

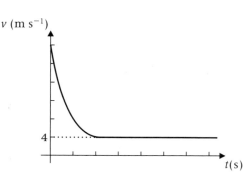

(c) To find when the speed of the parachutist drops to 5 m s^{-1}, we must solve the equation below.

$$5 = 26e^{-\frac{t}{2}} + 4$$

$$e^{-\frac{t}{2}} = \frac{1}{26}$$

$$-\frac{t}{2} = \ln\left(\frac{1}{26}\right)$$

$$t = 2\ln(26)$$

$$= 6.52 \text{ s}$$

Now this value of t can be substituted into the expression for s.

$$s = 52\left(1 - e^{-\frac{2\ln 26}{2}}\right) + 4 \times 2 \ln 26$$

$$= 76.1 \text{ m (to 3 sf)}$$

So the minimum safe height to open the parachute is 76.1 m.

EXERCISE 4B

1 The acceleration, a m s^{-2}, at time t seconds of a particle that starts at rest is given by $a = \dfrac{t^2}{100}$.

(a) Find an expression for the velocity of the particle at time t.

(b) Find the velocity of the particle when $t = 5$.

(c) Find the distance that the particle travels in the first 5 seconds of its motion.

2 The acceleration of a cyclist, at time t seconds is given by $a = 2 - \dfrac{t}{10}$ m s^{-2}. This model is valid until $t = 8$ s. If the cyclist starts at rest, find the distance travelled in the 8 seconds and the final speed of the cyclist.

3 A body experiences an acceleration of $0.1t$ m s^{-2}, at time t seconds, for $0 \leq t \leq 5$. Its acceleration is zero for $t > 5$. Find the distance travelled by the body when $t = 10$, if it has an initial velocity of 3 m s^{-1}.

4 A train travels along a straight set of tracks. Initially it moves with velocity 10 m s^{-1}. It then experiences an acceleration given by $a = 0.1t(4 - t)$ m s^{-2}, until $t = 4$ s. Find the velocity when $t = 4$ and the distance travelled at this time.

5 The graph below shows how the magnitude of the force exerted on a lorry, of mass 10 000 kg, by its brakes varies with time. The lorry initially has a velocity of 20 m s^{-1}.

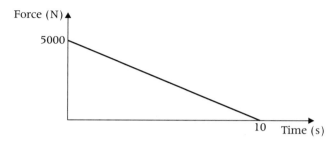

   (a) Show that the acceleration, a m s^{-2}, at time t seconds is given by $a = \dfrac{t}{20} - \dfrac{1}{2}$ for $0 \leqslant t \leqslant 10$.

   (b) Find the velocity of the lorry when $t = 10$.

   (c) Find the distance travelled by the lorry in the 10 second period.

6 The acceleration of a car at time t seconds is $\dfrac{30 - t}{10}$ m s^{-2}.

The car starts at rest.

   (a) Find an expression for the velocity of the car at time t.

   (b) When is the acceleration of the car zero and what is its speed at this time?

   (c) Find the total distance travelled by the car in the first 30 seconds of its motion.

7 A particle moves along a straight line. At time t seconds the acceleration, a m s^{-2}, of the particle is given by $a = 2 - 2\mathrm{e}^{-t}$.

At time $t = 0$ the particle is at the origin moving with a velocity of 4 m s^{-1}.

   (a) Show that the velocity, v m s^{-1}, at time t seconds is given by $v = 2t + 2\mathrm{e}^{-t} + 2$.

   (b) Find an expression for the distance of the particle from the origin at time t. [A]

8 A mass that is attached to one end of a spring moves up and down. The velocity, v m s^{-1}, at time t seconds of the mass is given by $v = 0.6 \sin 3t$.

   (a) Find the acceleration of the mass at time t.

   (b) Find the displacement of the mass at time t if its initial displacement is 0.8 m from the fixed end of the spring.

   (c) Sketch a displacement–time graph for the mass.

9 The acceleration, a m s^{-2}, at time t seconds of a stone that falls from rest is modelled as $9.8e^{-2t}$.

(a) Find expressions for the velocity and displacement of the stone at time t.

(b) Sketch a velocity–time graph for the stone.

(c) How far would the stone fall in 20 seconds?

10 A bullet is fired horizontally into a fluid, with an initial velocity of 40 m s^{-1}. Its acceleration at time t seconds is modelled as $-50e^{-0.5t}$ m s^{-2}. This model only applies while the bullet is moving.
Find the distance that the bullet travels into the fluid. Assume that the bullet always moves horizontally.

11 The tip of the blade in an electric jigsaw moves so that its acceleration is $40\cos(100\pi t)$ m s^{-2}, t seconds after it starts to move from rest at its lowest position.

(a) Find the velocity of the tip of the blade at time t.

(b) What is the maximum velocity of the tip of the blade?

(c) What is the maximum displacement of the tip of the blade from its lowest position?

12 A particle moves so that its acceleration at time t seconds is $(4\sin 2t + 6\cos 2t)$ m s^{-2}. Initially the particle has a velocity of -2 m s^{-1} and is at the origin.

(a) Find an expression for the position of the particle at time t.

(b) Determine the maximum speed of the particle.

4.4 Motion in two or three dimensions

The ideas that you have used in one dimension can be very easily extended to two and three dimensions. A position vector would be written as $\mathbf{r} = x\mathbf{i} + x\mathbf{j} + z\mathbf{k}$ in three dimensions or as $\mathbf{r} = x\mathbf{i} + y\mathbf{j}$ in two dimensions. Here x, y and z are all functions of time. To obtain velocities and accelerations we simply need to differentiate these with respect to time. So in three dimensions we have

$$\mathbf{r} = x\mathbf{i} + y\mathbf{j} + z\mathbf{k}$$

$$\mathbf{v} = \frac{d\mathbf{r}}{dt}$$

$$= \frac{dx}{dt}\mathbf{i} + \frac{dy}{dt}\mathbf{j} + \frac{dz}{dt}\mathbf{k}$$

$$\mathbf{a} = \frac{d\mathbf{v}}{dt}$$

$$= \frac{d^2x}{dt^2}\mathbf{i} + \frac{d^2y}{dt^2}\mathbf{j} + \frac{d^2z}{dt^2}\mathbf{k}$$

In two dimensions these results reduce to

$$\mathbf{r} = x\mathbf{i} + y\mathbf{j}$$

$$\mathbf{v} = \frac{d\mathbf{r}}{dt}$$

$$= \frac{dx}{dt}\mathbf{i} + \frac{dy}{dt}\mathbf{j}$$

$$\mathbf{a} = \frac{d\mathbf{v}}{dt}$$

$$= \frac{d^2x}{dt^2}\mathbf{i} + \frac{d^2y}{dt^2}\mathbf{j}$$

These results are used in the following examples.

Worked example 4.7

The position vector, $\mathbf{r}$ m, at time t seconds of an aeroplane that is circling an airport is given by

$$\mathbf{r} = 500 \sin\left(\frac{t}{20}\right)\mathbf{i} + 500 \cos\left(\frac{t}{20}\right)\mathbf{j} + 4000\mathbf{k}$$

The unit vectors $\mathbf{i}$ and $\mathbf{j}$ are east and north, respectively, and $\mathbf{k}$ is vertical.

(a) Find the velocity of the aeroplane.

(b) Find the magnitude of the acceleration of the aeroplane.

Solution

(a) The velocity can be found by differentiating the position vector with respect to time.

$$\mathbf{v} = \frac{d\mathbf{r}}{dt}$$

$$= \frac{d}{dt}\left(500 \sin\left(\frac{t}{20}\right)\right)\mathbf{i} + \frac{d}{dt}\left(500 \cos\left(\frac{t}{20}\right)\right)\mathbf{j} + \frac{d}{dt}(4000)\mathbf{k}$$

$$= 25 \cos\left(\frac{t}{20}\right)\mathbf{i} - 25 \sin\left(\frac{t}{20}\right)\mathbf{j}$$

(b) The velocity can be differentiated to obtain the acceleration.

$$\mathbf{a} = \frac{d\mathbf{v}}{dt}$$

$$= \frac{d}{dt}\left(25 \cos\left(\frac{t}{20}\right)\right)\mathbf{i} + \frac{d}{dt}\left(-25 \sin\left(\frac{t}{20}\right)\right)\mathbf{j}$$

$$= -\frac{5}{4}\sin\left(\frac{t}{20}\right)\mathbf{i} - \frac{5}{4}\cos\left(\frac{t}{20}\right)\mathbf{j}$$

Now consider the magnitude of the acceleration.

$$a = \sqrt{\left(-\frac{5}{4}\sin\left(\frac{t}{20}\right)\right)^2 + \left(-\frac{5}{4}\cos\left(\frac{t}{20}\right)\right)^2}$$

$$= \frac{5}{4}\sqrt{\sin^2\left(\frac{t}{20}\right) + \cos^2\left(\frac{t}{20}\right)}$$

$$= \frac{5}{4} \text{ m s}^{-2}$$

> Remember that $\sin^2\theta + \cos^2\theta \equiv 1$.

Obtaining a position vector

In the same way that accelerations and velocities were integrated in one dimension, you can integrate velocity or acceleration vectors to obtain velocities or displacements. In three dimensions we would integrate the acceleration to obtain the velocity.

$$\mathbf{a} = a_x\mathbf{i} + a_y\mathbf{j} + a_z\mathbf{k}$$

$$\mathbf{v} = \int \mathbf{a} \, dt$$

$$= \int a_x \, dt \, \mathbf{i} + \int a_y \, dt \, \mathbf{j} + \int a_z \, dt \, \mathbf{k}$$

Similarly we would integrate the velocity to get a position vector.

$$\mathbf{v} = v_x\mathbf{i} + v_y\mathbf{j} + v_z\mathbf{k}$$

$$\mathbf{r} = \int \mathbf{v} \, dt$$

$$= \int v_x \, dt \, \mathbf{i} + \int v_y \, dt \, \mathbf{j} + \int v_z \, dt \, \mathbf{k}$$

Note that when integrating like this you will introduce a number of constants of integration. It is important to determine each of these using the initial velocity and initial position or other similar information.

Worked example 4.8

A particle moves so that its acceleration, $\mathbf{a}$ m s^{-2}, at time t seconds is given by:

$$\mathbf{a} = 0.6t\mathbf{i} + (1 - 1.2t)\mathbf{j}$$

where $\mathbf{i}$ and $\mathbf{j}$ are perpendicular unit vectors.

(a) Find the velocity of the particle at time t if it has an initial velocity of $(2\mathbf{i} + 3\mathbf{j})$ m s^{-1}.

(b) Find an expression for the position of the particle at time t if its initial position is $20\mathbf{i}$ metres.

(c) Find the speed of the particle when $t = 2$.

Solution

(a) To find the velocity at time t integrate the acceleration vector:

$$\mathbf{v} = \int 0.6t \, dt \, \mathbf{i} + \int (1 - 1.2t) \, dt \, \mathbf{j}$$
$$= (0.3t^2 + c_1)\mathbf{i} + (t - 0.6t^2 + c_2)\mathbf{j}$$

Using the initial velocity, $2\mathbf{i} + 3\mathbf{j}$, gives $c_1 = 2$ and $c_2 = 3$, so that the velocity is

$$\mathbf{v} = (0.3t^2 + 2)\mathbf{i} + (t - 0.6t^2 + 3)\mathbf{j}.$$

(b) To find the position at time t integrate the velocity vector with respect to t:

$$\mathbf{r} = \int (0.3t^2 + 2) \, dt \, \mathbf{i} + \int (t - 0.6t^2 + 3) \, dt \, \mathbf{j}$$
$$= (0.1t^3 + 2t + c_3)\mathbf{i} + (0.5t^2 - 0.2t^3 + 3t + c_4)\mathbf{j}$$

Using the initial position of $20\mathbf{i}$ gives $c_3 = 20$ and $c_4 = 0$, so that the position is given by

$$\mathbf{r} = (0.1t^3 + 2t + 20)\mathbf{i} + (0.5t^2 - 0.2t^3 + 3t)\mathbf{j}.$$

(c) When $t = 2$:

$$\mathbf{v} = (0.3 \times 2^2 + 2)\mathbf{i} + (2 - 0.6 \times 2^2 + 3)\mathbf{j}$$
$$= 3.2\mathbf{i} + 2.6\mathbf{j}$$

Then the speed v is given by:

$$v = \sqrt{3.2^2 + 2.6^2}$$
$$= 4.12 \text{ m s}^{-1}$$

EXERCISE 4C

1 A particle moves so that, at time t seconds, its position vector in metres is given by

$$\mathbf{r} = (t^2 - 5)\mathbf{i} + (4 - t + 6t^2)\mathbf{j}.$$

  (a) Find the velocity and acceleration of the particle at time t.
  (b) Find the position and velocity of the particle when $t = 4$.

2 Two aeroplanes, A and B, move so that at time t seconds their position vectors, in metres, are given by

$$\mathbf{r}_A = (30t - 600)\mathbf{i} + (3t^2 - 120t + 1400)\mathbf{j}$$

and

$$\mathbf{r}_B = (20t + 10)\mathbf{i} + (40t - 10)\mathbf{j},$$

where $\mathbf{i}$ and $\mathbf{j}$ are unit vectors that are directed east and north, respectively.

  (a) Find the velocities of A and B at time t.
  (b) Find the speed of B.
  (c) Find the time when the two aeroplanes are travelling in parallel directions and the distance between them at this time.

3 A ball rolls on a slope so that its position is given by $\mathbf{r} = (t^2\mathbf{i} + 2t\mathbf{j})$ m at time t seconds, where $\mathbf{i}$ and $\mathbf{j}$ are perpendicular unit vectors. Find the velocity and acceleration of the ball at time t.

4 A light aircraft moves so that its position, in metres, relative to an origin O, at time t seconds is given by

$$\mathbf{r} = \left(4t - \frac{t^2}{5}\right)\mathbf{i} + 10t\mathbf{j},$$

where $\mathbf{i}$ and $\mathbf{j}$ are unit vectors that are directed east and north, respectively.

(a) Find an expression for the velocity of the aircraft at time t.

(b) Find the time when the aircraft is due north of its initial position, and the distance from its initial position at that time.

(c) Find the time when the aircraft is travelling north and its speed at that time.

(d) Describe fully the acceleration of the aircraft.

5 The position, in metres, of a particle at time t seconds is given by

$$\mathbf{r} = (t^2 - 8t + 2)\mathbf{i} + (2t^3 - 5t^2 + 6t)\mathbf{j},$$

where $\mathbf{i}$ and $\mathbf{j}$ are horizontal and vertical unit vectors, respectively. The mass of the particle is 3 kg.

(a) Find an expression for the velocity of the particle at time t.

(b) Find an expression for the resultant force acting on the particle at time t.

(c) Find when the horizontal component of the velocity is zero, and the position of the particle at this time.

6 The position, at time t seconds, of a car overtaking a lorry is modelled, in metres, as $\mathbf{r} = 20t\mathbf{i} + 5\left(t - \frac{t^2}{10}\right)\mathbf{j}$, where $\mathbf{i}$ and $\mathbf{j}$ are unit vectors parallel and perpendicular to the straight path of the lorry. The lorry travels along a straight line and has position, in metres, given by $\mathbf{r} = (10 + 15t)\mathbf{i}$ at time t seconds. Both the car and the lorry are modelled as particles.

(a) Find the time when the car is level with the lorry and its speed at this time.

(b) Find the time when the car is travelling parallel to the lorry and its acceleration at this time.

7 A force of magnitude $5t\mathbf{i} + 10t\mathbf{j}$ N acts on a body, of mass 50 kg, at time t seconds, for $0 \leq t \leq 5$. No force acts on the body for $t > 5$. Find the displacement of the body when $t = 10$, if it has an initial velocity of $3\mathbf{i}$ m s^{-1} and starts at the origin. The unit vectors $\mathbf{i}$ and $\mathbf{j}$ are perpendicular.

8 A particle moves so that its velocity, in metres per second, at time t seconds is $\mathbf{v} = (4t^2 + 3)\mathbf{i} + (37.5 - 15t)\mathbf{j}$, where $\mathbf{i}$ and $\mathbf{j}$ are perpendicular unit vectors. Initially the particle is at the origin. When the particle is moving parallel to the unit vector $\mathbf{i}$ the magnitude of the resultant force acting on the particle is 80 N.
  (a) Find the mass of the particle.
  (b) Find the position of the particle when it is moving parallel to the unit vector $\mathbf{i}$.

9 A jet ski has an initial velocity of $(2\mathbf{i} + 5\mathbf{j})$ m s^{-1} and experiences an acceleration of $(\mathbf{i} + 0.2t\mathbf{j})$ m s^{-2}, at time t seconds. Find expressions for the velocity and position of the jet ski at time t seconds. Assume that the jet ski starts at the origin.

10 The acceleration of a particle at time t seconds is $(2t\mathbf{i} - 5t\mathbf{j})$ m s^{-2}. Initially the particle is at the origin and has velocity $(3\mathbf{i} + 6\mathbf{j})$ m s^{-1}.
  (a) Find an expression for the velocity of the particle at time t.
  (b) Find an expression for the position of the particle at time t.

11 The acceleration of a particle at time t seconds is $(4\mathbf{i} - t\mathbf{j})$ m s^{-2}. The particle is initially at rest at the point with position vector $(5\mathbf{i} - 10\mathbf{j})$ m. Note that $\mathbf{i}$ and $\mathbf{j}$ are perpendicular unit vectors.
  (a) Find an expression for the velocity of the particle at time t.
  (b) Find an expression for the position of the particle at time t.

12 A particle, P, moves so that at time t seconds its position vector, $\mathbf{r}$ metres, is
$$\mathbf{r} = \begin{bmatrix} 2t^2 + 6 \\ 5t \end{bmatrix}, \quad 0 \leq t \leq 5.$$
  (a) Find the velocity of P at time t.
  (b) The force acting on P is $\begin{bmatrix} 2 \\ 0 \end{bmatrix}$ newtons. Find the mass of P.
  (c) At the instant when $t = 5$, an additional force, $\begin{bmatrix} 0 \\ t \end{bmatrix}$ newtons, begins to act on P.
    (i) Find the resultant acceleration of P.
    (ii) Find the velocity of P when $t = 10$. [A]

13 At time t, the position vector of a particle Q is
$$\mathbf{r} = (t^2 - 6t + 4)\mathbf{i} + \tfrac{1}{3}t^3\mathbf{j}.$$

(a) Find the velocity of Q at time t.

(b) Find the value of t when Q is moving parallel to the vector $\mathbf{j}$.

(c) Find the acceleration of Q and state, with a reason, whether or not this acceleration is constant. [A]

14 A particle has mass 2000 kg. A single force, $\mathbf{F} = 1000t\mathbf{i} - 5000\mathbf{j}$ newtons, acts on the particle, at time t seconds. The unit vectors $\mathbf{i}$ and $\mathbf{j}$ are perpendicular. No other forces act on the particle.

(a) Find an expression for the acceleration of the particle.

(b) A time $t = 0$, the velocity of the particle is $6\mathbf{j}$ m s^{-1}. Show that at time t the velocity, $\mathbf{v}$ m s^{-1}, of the particle is given by
$$\mathbf{v} = \frac{t^2}{4}\mathbf{i} + \left(6 - \frac{5t}{2}\right)\mathbf{j}.$$

(c) The particle is initially at the origin. Find an expression for the position vector, $\mathbf{r}$ metres, of the particle at time t seconds. [A]

15 A particle moves so that at time t seconds, its position, $\mathbf{r}$ metres, is given by
$$\mathbf{r} = (t^3 - 3t^2)\mathbf{i} + (4t + 2t^2)\mathbf{j},$$
where $\mathbf{i}$ and $\mathbf{j}$ are perpendicular unit vectors.

(a) By differentiating, find the velocity of the particle at time t.

(b) Find, but do not simplify, an expression for the magnitude of the acceleration of the particle.

(c) Find the time when the magnitude of the acceleration is a minimum and find its magnitude at this time. [A]

16 A particle moves, so that its acceleration at time t, is given by
$$\mathbf{a} = -4\cos t\,\mathbf{i} + 3\sin t\,\mathbf{j} + \tfrac{1}{2}\mathbf{k},$$
where the unit vectors $\mathbf{i}$, $\mathbf{j}$ and $\mathbf{k}$ are mutually perpendicular. The initial position of the particle is $4\mathbf{i}$ and its initial velocity is $\tfrac{1}{2}\mathbf{j}$.

(a) Find an expression for the velocity of the particle at time t.

(b) Find the position vector of the particle at time t.

(c) Find the distance of the particle from the origin when $t = \dfrac{\pi}{2}$. [A]

17 A boat moves so that its position vector, **r** metres, at time t seconds is
$$\mathbf{r} = 40\cos\left(\frac{t}{20}\right)\mathbf{i} + 80\sin\left(\frac{t}{20}\right)\mathbf{j}.$$
The unit vectors **i** and **j** are directed east and north, respectively.

(a) Find an expression for the velocity of the boat at time t.

(b) In what direction is the boat travelling when $t = 0$? Justify your answer.

(c) At what time is the boat travelling due south for the first time?

18 An object that describes a circular path has a position vector, in metres, at time t seconds given by $\mathbf{r} = 4\sin(8t)\mathbf{i} + 4\cos(8t)\mathbf{j}$. Find the magnitude of the velocity and acceleration of the object.

19 A particle follows a path so that its position at time t is given by
$$\mathbf{r} = 4\cos(2t)\mathbf{i} + 3\sin(2t)\mathbf{j}.$$

(a) Find the position, velocity and acceleration of the particle when $t = \dfrac{\pi}{2}$.

(b) Show that the magnitude of the acceleration at time t is
$$\sqrt{(144 + 112\cos^2(2t))}$$
and find its maximum and minimum values. [A]

20 A sky diver jumps at time $t = 0$ from an aeroplane that is travelling horizontally. The velocity, **v** m s^{-1}, of the sky diver at time t seconds is given by
$$\mathbf{v} = 70\,e^{-0.1t}\mathbf{i} + 40(e^{-0.1t} - 1)\mathbf{j},$$
where **i** and **j** are unit vectors in the horizontal and upward vertical directions respectively.

(a) Describe what happens to the velocity of the sky diver as t increases.

(b) Taking the origin to be the initial position of the sky diver, find an expression for his position vector at time t seconds. [A]

21 A particle moves in a horizontal plane and its position vector at time t is relative to a fixed origin O is given by
$$\mathbf{r} = (2\sin t\,\mathbf{i} + \cos t\,\mathbf{j})\text{ m}.$$
Find the values of t in the range $0 \leq t \leq \pi$ when the speed of the particle is a maximum. [A]

Key point summary

1. In one, two or three dimensions displacements or position vectors can be differentiated to give velocities and accelerations. *pp69, 81*

2. Accelerations can be integrated to give velocities and position vectors or displacements. *pp75, 81*

Formulae to learn

$$v = \frac{dx}{dt}$$

$$a = \frac{dv}{dt}$$

$$a = \frac{d^2x}{dt^2}$$

$$v = \int a \, dt$$

$$s = \int v \, dt$$

$$\mathbf{r} = x\mathbf{i} + y\mathbf{j} + z\mathbf{k}$$

$$\mathbf{v} = \frac{d\mathbf{r}}{dt}$$

$$= \frac{dx}{dt}\mathbf{i} + \frac{dy}{dt}\mathbf{j} + \frac{dz}{dt}\mathbf{k}$$

$$\mathbf{a} = \frac{d\mathbf{v}}{dt}$$

$$= \frac{d^2x}{dt^2}\mathbf{i} + \frac{d^2y}{dt^2}\mathbf{j} + \frac{d^2z}{dt^2}\mathbf{k}$$

$$\mathbf{v} = \int \mathbf{a} \, dt$$

$$\mathbf{r} = \int \mathbf{v} \, dt$$

Test yourself

1 The displacement, s metres, of a particle at time t seconds is given by $s = 5t + 6e^{-2t} - 4$.
  (a) Find the initial velocity of the particle.
  (b) Sketch a velocity–time graph for the particle.
  (c) Find the acceleration at time t seconds.

What to review: Section 4.2

2 A particle moves so that at time t seconds its acceleration is $-\left(\dfrac{\sqrt{t}}{2}\right)$ m s^{-2}. Its initial velocity is 9 m s^{-1}, when it passes the origin.
Find the displacement of the particle from the origin when it comes to rest.

Section 4.3

3 A particle moves so that its position vector at time t is given by
$$\mathbf{r} = (6 + 2t)\mathbf{i} + (8 - 3t^2)\mathbf{j},$$
where $\mathbf{i}$ and $\mathbf{j}$ are perpendicular unit vectors.
  (a) Find the position of the particle when $t = 10$.
  (b) Find the velocity of the particle in terms of t.
  (c) Find the speed of the particle when $t = 2$.
  (d) Find the acceleration of the particle.

Section 4.4

4 A particle moves so that its acceleration, $\mathbf{a}$ m s^{-2}, at time t seconds is given by
$$\mathbf{a} = 0.6t\mathbf{i} + (1 - 1.2t)\mathbf{j},$$
where $\mathbf{i}$ and $\mathbf{j}$ are perpendicular unit vectors.
  (a) Find the velocity of the particle at time t if it has an initial velocity of $(2\mathbf{i} + 3\mathbf{j})$ m s^{-1}.
  (b) Find an expression for the position of the particle at time t if its initial position is $20\mathbf{i}$.
  (c) Find the speed of the particle when $t = 2$.

Section 4.4

Test yourself ANSWERS

1 (a) -7 m s^{-1} (c) $a = 24e^{-2t}$.

2 48.6 m.

3 (a) $26\mathbf{i} - 292\mathbf{j}$; (b) $2\mathbf{i} - 6t\mathbf{j}$; (c) 12.2 m s^{-1}; (d) $-6\mathbf{j}$.

4 (a) $\mathbf{v} = (2 + 0.3t^2)\mathbf{i} + (3 + t - 0.6t^2)\mathbf{j}$;
  (b) $\mathbf{r} = (20 + 2t + 0.1t^3)\mathbf{i} + (3t + 0.5t^2 - 0.2t^3)\mathbf{j}$;
  (c) 4.12 m s^{-1}.

CHAPTER 5
Circular motion

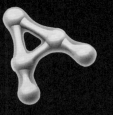

Learning objectives

After studying this chapter you should be able to:
- convert from rpm to rad s^{-1}
- know that the velocity is directed along the tangent of the circle
- know that the acceleration is directed towards the centre of the circle
- know that $v = r\omega$
- know that $a = r\omega^2 = \dfrac{v^2}{r}$
- solve problems for motion in horizontal circles at constant speeds.

5.1 Introduction

This chapter will focus on the motion of objects that travel in a circle with constant speed. There are many everyday situations which can be modelled in this way, for example, fairground rides, satellites orbiting the Earth and clothes in a spin-dryer. Also cars travelling around corners or on roundabouts travel round part of a circle. To analyse such situations we need to apply Newton's second law after finding the resultant force along with the acceleration of the object travelling round the circle. Before doing this you must be familiar with the concept of angular speed.

5.2 Angular speed

Imagine a line drawn from the centre of the circle to the object that is travelling round the circle. The rate at which this line rotates about the centre is called the angular speed of the object. This could be given in terms of rpm (revolutions per minute), as with car engines, but it is most often given in radians per second (rad s^{-1}).

We will now consider an example that makes use of angular speed. To do this you will need to recall that one complete revolution is equivalent to turning through 2π radians.

$$1 \text{ rpm} = \frac{2\pi}{60} \text{ rad s}^{-1}$$

Worked example 5.1

The angular speed of a turntable is given as $33\frac{1}{3}$ rpm. Find the angular speed of the turntable in radians per second.

Solution

In one revolution there are 2π radians.

So $33\frac{1}{3}$ rpm $= 33\frac{1}{3} \times 2\pi$ radians per minute

$$= \frac{33\frac{1}{3} \times 2\pi}{60} \text{ radians per second}$$

$$= \frac{10\pi}{9}$$

$$= 3.49 \text{ rad s}^{-1} \text{ (to 3 sf)}$$

Velocity and circular motion

Now consider a particle, which moves with angular speed ω rad s^{-1}. The angle turned through in 1 second will be ω radians, so the angle turned through in t seconds will be $\theta = \omega t$ radians. If the particle moves in a circle with radius r m, the distance moved by the particle will be the length of the arc AB.

$$AB = r\theta$$

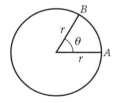

Each second the particle moves a distance $r\omega$, so the speed, v, of the particle is given by

$$v = r\omega.$$

The velocity has magnitude $r\omega$ and is directed along a tangent to the circle.

Acceleration and circular motion

Newton's first law of motion states that a particle will move in a straight line unless acted upon by a force. Therefore, if a particle moves in a circle, there must be a force acting upon it. If the speed is constant then the resultant force will have no component in the direction of motion. Hence the resultant force on the particle will be perpendicular to the velocity and must always act towards the centre. Consequently, if the resultant force is towards the centre, then so is the acceleration of the particle. You will now find the magnitude of this acceleration.

Suppose a particle moves in a circle with centre O, radius r, and with constant angular velocity ω.

Suppose, further, that the particle starts from A($t = 0$) so $\theta = \omega t$. The position vector of the particle **OP**, in terms of the unit vectors **i** and **j** will be given by

$$\begin{aligned}\mathbf{r} &= x\mathbf{i} + y\mathbf{j} \\ &= r\cos\theta\,\mathbf{i} + r\sin\theta\,\mathbf{j} \\ &= r\cos\omega t\,\mathbf{i} + r\sin\omega t\,\mathbf{j}.\end{aligned}$$

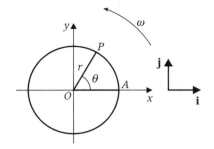

If you differentiate once you get the velocity vector and twice you get the acceleration vector.

$$\mathbf{v} = -\omega r\sin\omega t\,\mathbf{i} + \omega r\cos\omega t\,\mathbf{j}$$

and

$$\mathbf{a} = -\omega^2 r\cos\omega t\,\mathbf{i} - \omega^2 r\sin\omega t\,\mathbf{j}$$

The magnitude of the acceleration of the particle is

$$a = \omega^2 r\sqrt{\cos^2\omega t + \sin^2\omega t}.$$

But $\cos^2\theta + \sin^2\theta \equiv 1$, so

$$a = r\omega^2$$

Sometimes it is useful to express the magnitude of the acceleration in terms of the speed, v, rather than ω. From $v = r\omega$, note that $\omega = \dfrac{v}{r}$ and substitute this so that the expression for the magnitude of the acceleration becomes

$$a = r\left(\frac{v}{r}\right)^2 = \frac{v^2}{r}.$$

To confirm that the direction of the particle's acceleration is towards the centre compare the acceleration and position vectors.

$$\mathbf{a} = -\omega^2(r\cos\omega t\,\mathbf{i} + r\sin\omega t\,\mathbf{j})$$
$$\mathbf{a} = -\omega^2 \mathbf{r}$$

The position vector is directed outwards from the centre of the circle. The acceleration is in the opposite direction and so is directed **towards** the centre.

Worked example 5.2

If the radius of the Earth is taken as 6370 km find the speed and magnitude of the acceleration of a man standing on the Equator.

$$\omega = 1 \text{ rev per day}$$
$$= \frac{2\pi}{24 \times 60 \times 60} = 7.27 \times 10^{-5} \text{ rad s}^{-1}$$
$$v = 6\,370\,000 \times 7.27 \times 10^{-5} = 463 \text{ m s}^{-1}$$
$$a = 6\,370\,000 \times (7.27 \times 10^{-5})^2 = 3.37 \times 10^{-2} \text{ m s}^{-2}$$

EXERCISE 5A

1. What is the angular speed of the minute hand of a clock in
   (a) revolutions per minute, (b) radians per second.

2. A wheel makes 100 revolutions in 10 minutes. Find its angular speed in radians per second.

3. The distance of a satellite from the centre of a planet is approximately 50 000 km. Estimate the speed of the satellite relative to the planet in metres per second. (Assume that the satellite orbits the planet in 12 hours in a circular path.)

4. The distance from the Sun to the Earth is approximately 150×10^6 km. Estimate the speed of the Earth relative to the Sun. (Assume that the orbit of the Earth is a circle and that the Earth orbits the Sun once every year.)

5. Find the speed and the magnitude of the acceleration of a particle, which moves in a circle with radius 20 cm, and angular speed of 2500 rpm.

6. Find the magnitude of the resultant force on a particle of mass 250 grams, which moves in a circle of radius 10 m and angular speed of 36 rpm.

7. A particle moves in a circle with speed 5 m s^{-1}. Given that the acceleration has magnitude 10 m s^{-2}, find the radius of the circle.

8. What is the magnitude of the acceleration of a truck, which goes round a bend of radius 20 m at a speed of 20 km h^{-1}.

9. A washing machine spins at 1000 rpm. The drum has diameter 40 cm. What are the speed and magnitude of the acceleration of a sock, which is stuck to the edge of the drum during the spinning cycle?

10. Joe and Tom ride on a fairground roundabout. Joe is 2 m and Tom is 1.5 m from the centre of rotation and the roundabout is rotating at 10 rpm. Find:
    (a) the angular speed of the roundabout in radians per second,
    (b) the speeds of Joe and Tom in metres per second.

11. The orbit of the Moon around the Earth may be modelled as circular. The time taken for the Moon to make one complete orbit is approximately 27.32 days.
    (a) Show that the angular speed of the Moon is approximately 2.66×10^{-6} radians per second.
    (b) Assuming that the radius of the circular orbit is approximately 3.844×10^8 m, find the speed of the Moon, relative to the Earth, in metres per second. [A]

12

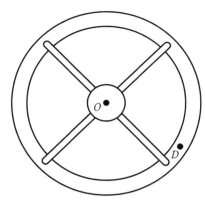

A children's roundabout has a horizontal, circular base with centre O. Stephanie places a doll, D, on the roundabout and then pushes the roundabout so that it rotates with constant angular speed ω radians per second. Stephanie notices that the doll makes 8 complete revolutions about O in one minute.

(a) Show that $\omega = \dfrac{4\pi}{15}$.

(b) The doll's speed is 0.75 m s^{-1}. Find its distance from O, giving your answer to two significant figures.

(c) Find the acceleration of the doll and indicate its direction on a diagram. [A]

5.3 Forces involved in horizontal circular motion

In the previous section you saw that when a particle, of mass m, moves in a circle with constant speed, its acceleration is of magnitude $r\omega^2$ (or v^2/r) and acts towards the centre of the circle (where r is the radius of the circle, v is the speed of the particle and ω is its angular speed). The resultant force, therefore, must always act towards the centre. Newton's second law implies that the magnitude of the resultant force will be given by

$$F = mr\omega^2 \qquad \text{(or } mv^2/r\text{)}.$$

If the circle of motion is horizontal then any vertical components of force must cancel.

> When a particle moves in a horizontal circle with constant speed there are two principles that can always be applied:
>
> - the resultant of vertical components of forces must be zero;
> - $F = ma$ can be applied radially.

Worked example 5.3

A particle P, of mass 100 grams, is attached to one end of a light inextensible string of length 50 cm. The other end of the string is fixed at O, on a smooth horizontal surface. The particle moves in a circle on the horizontal surface, with centre O, at 180 rpm.

(a) Find the normal reaction on the particle.

(b) Find the tension in the string.

Solution

The diagram shows the forces acting on the particle, its weight, the normal reaction from the surface and the tension in the string.

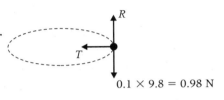

(a) The acceleration of P is horizontal, so the resultant of the vertical forces must be zero.

$R = 0.98$ N

(b) The angular speed must be converted from revolutions per minute to radians per second.

$\omega = 180$ rpm

$= 180 \times \dfrac{2\pi}{60}$ rad s^{-1}

$= 6\pi$ rad s^{-1}

Using $F = ma$ in the radial direction with $a = r\omega^2$ gives

$T = 0.1 \times 0.5 \times (6\pi)^2$
$= 17.8$ N

Worked example 5.4

A turntable rotates at $33\tfrac{1}{3}$ rpm. A counter, of mass 10 grams, is placed 10 cm from the centre of the circle and does not slip. Find the minimum possible value of the coefficient of friction.

Solution

The diagram shows the forces acting on the counter.

First express the angular speed in radians per second.

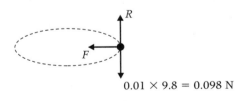

$\omega = \dfrac{33\tfrac{1}{3} \times 2\pi}{60}$

$= \dfrac{10\pi}{9}$ rad s^{-1}

Using $F = ma$ radially with $a = r\omega^2$

$F = 0.01 \times 0.1 \left(\dfrac{10\pi}{9}\right)^2$

$= 0.012\,18$ N

Resolving vertically

$$R = 0.01 \times 9.8 = 0.098 \text{ N}$$

Now we can use the friction inequality $F \leq \mu R$, to give

$$0.012\,18 \leq \mu \times 0.098$$
$$\mu \geq 0.124 \text{ (to 3 sf)}$$

So the least value of μ is 0.124.

EXERCISE 5B

1 A particle, of mass 2 kg, is attached to a fixed point on a smooth horizontal table by a light inextensible string of length 50 cm. The particle travels in a circle on the table at 400 rpm. Find the tension in the string.

2 A particle, of mass 2 kg, is moving in a circle of radius 5 m with a constant speed of 3 m s^{-1}. What is the magnitude and direction of the resultant force acting on the particle?

3 A particle, of mass 3 kg, moves in a circle on a smooth horizontal plane. The particle is attached to a fixed point, O, on the plane, by a light inextensible string of length 1.5 m. If the velocity of the particle is 6 m s^{-1}, find the tension in the string.

4 An inextensible string has length 3 m. It is fixed at one end to a point O on a smooth horizontal table. A particle of mass 2 kg is attached to the other end and describes circles on the table with O as centre and the string taut. If the string breaks when the tension is 90 N, what is the maximum safe speed of the particle?

5 An athlete throwing the hammer swings it in preparation for his throw. Assume that the hammer travels in a horizontal circle of radius 2.0 m and is rotating at 1 revolution per second. The mass of the wire is negligible and the mass of the hammer is 7.3 kg. What is the tension in the wire attached to the hammer? (Ignore the weight of the hammer.)

6 A particle moves with constant angular speed around a circle of radius a and centre O. The only force acting on the particle is directed towards O and is of magnitude $\dfrac{k}{a^2}$ per unit mass, where k is a constant. Find, in terms of k and a, the time taken for the particle to make one complete revolution.

7 A satellite, of mass 1 tonne, orbits a planet every 24 hours. The satellite is at a height of 700 km above the surface of the planet. The radius of the planet is 6500 km. Find the force of gravity on the satellite.

8 A car, of mass 1 tonne, takes a bend, of radius 150 m, on a level road, at 80 km h^{-1}, without sliding. Find the frictional force between the tyres and the ground. What is the least value of μ?

9 Two particles, P and Q, are connected by a light inextensible string, that passes through a hole in a smooth horizontal table. Particle P has mass m and travels in a circle, whereas Q hangs in equilibrium under the table and has mass $2m$. If the radius of the circle in which P moves is 50 cm, find the angular velocity of P.

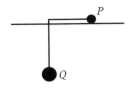

10 A horizontal turntable rotates at 45 rpm. A coin, of mass 50 grams, is placed on the turntable, at a distance 20 cm from the centre of rotation. The coefficient of friction between the coin and the turntable is 0.1.
   (a) Describe what happens to the coin.
   (b) What is the greatest distance from the centre of rotation that the coin can be placed without slipping?
   (c) How would your answers to (b) and (c) change for a heavier coin?

11 A penny, of mass m kg, is placed on a turntable 0.1 m from the centre. The turntable rotates at 45 rpm. If the penny is on the point of slipping, calculate the coefficient of friction between the penny and the turntable. Calculate the resultant force acting on the penny, when the turntable is rotating at 33 rpm, with the penny in the same position.

12 The position vector $\mathbf{r}$ (metres), of a particle P at time t seconds, is given by
   $$\mathbf{r} = 2\cos 3t\,\mathbf{i} + 2\sin 3t\,\mathbf{j},$$
   where unit vectors $\mathbf{i}$ and $\mathbf{j}$ are perpendicular. The mass of the particle is 4 kg.
   (a) Show that the particle moves with constant speed.
   (b) Find the speed of P.
   (c) State the angular speed of P.
   (d) Find the resultant force on P.

13 A car travels without slipping at 10 m s^{-1} around a horizontal bend of radius 30 m. Find the least value of the coefficient of friction.

14 The coefficient of friction between a road surface and the tyres on a car is 0.9. A horizontal bend on the road has a radius of 40 m. Find the maximum speed that the car can take the bend without sliding.

15 A fairground ride consists of a rotating cylinder. People stand on the inside of the cylinder with their backs to it. When the speed of rotation is great enough the floor is lowered so that only friction stops them from falling. The diameter of the cylinder is 10 m. The coefficient of friction between a person's body and the cylinder is taken as 0.5. Find the least angular speed necessary to stop a person from falling.

16 A child sits on a roundabout at a distance of 5 m from the centre of rotation and at a height of 2 m above ground level. The roundabout completes one revolution every 2 seconds. After one revolution the child drops a small toy, of mass 500 grams, that she was carrying, which then falls to the ground without hitting the roundabout.
   (a) Find the acceleration of the child on the roundabout and the magnitude of the force exerted on the toy by the child, before she drops it. Hint: include the weight of the toy.
   (b) Find the time that it takes for the toy to fall to the ground.
   (c) Sketch a graph to show how the magnitude of the acceleration of the toy varies with time. Assume $t = 0$ one revolution before the toy is dropped. [A]

17 A coin, of mass 0.01 kg, is placed on a horizontal turntable. The coin is at a distance of 5 cm from the centre of the turntable. The coefficient of friction between the coin and the turntable is 0.4. The turntable rotates about its centre, so that the coin follows a circular path at a constant speed, without slipping.
   (a) Calculate the maximum magnitude of the friction force acting on the coin.
   (b) Find the maximum angular speed of the turntable in revolutions per minute.
   (c) The angular speed of the turntable is halved. What happens to the magnitude of the friction force acting on the coin [A]

18 A strip of smooth metal, in the shape of a semicircle of radius 20 cm is fixed on a smooth horizontal surface. A marble, of mass 20 grams, is fired into the semicircle and travels at a speed of 5 m s^{-1}. Part of the path of the marble is shown by the dashed line in the diagram.
   (a) Find the magnitude of the acceleration of the marble in metres per second squared.
   (b) Show that the magnitude of the resultant of the reaction forces acting on the marble is approximately 2.51 N. Hint: include the weight of the marble.
   (c) Copy the diagram and show the path of the marble when it leaves the semicircle. [A]

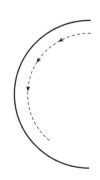

19 A device in a funfair consists of a hollow circular cylinder of radius 3 m, with a horizontal floor and vertical sides. A small child stands inside the cylinder and against the vertical side. The cylinder is rotated about its vertical axis of symmetry.

When the cylinder is rotating at a steady angular speed of 30 rpm the floor of the cylinder is lowered, so that the child is in contact only with the vertical side.

Given that the child does not slip, find, correct to two decimal places, the minimum coefficient of friction between the child and the side. [A]

20 A traffic roundabout has a radius of 80 m. The road surface at the roundabout is horizontal.

(a) Find the magnitude of the resultant force that acts on a car, of mass 1200 kg, travelling round the roundabout at 20 m s^{-1}.

Assume that the only horizontal force that acts on the car is friction. The coefficient of friction between the tyres and the road is μ.

(b) Find an inequality that μ must satisfy for the car to follow a circular path round the roundabout.

Typical values of μ for road vehicles are between 0.6 and 0.8.

(c) Calculate a safe speed limit for the roundabout.

(d) What factor have your omitted from your calculation that would have reduced the speed limit found in **(c)**? [A]

21 Two particles are connected by a light, inextensible string. One particle, which has mass 5 kg, follows a circular path of radius 70 cm, on a smooth horizontal table. The string passes through a hole in the table and a second particle, of mass 10 kg, hangs in equilibrium on the other end of the string as shown in the diagram.

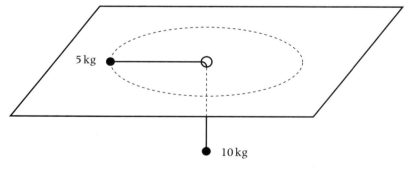

(a) By considering the 10 kg particle, calculate the tension in the string.

(b) Find the speed of the particle on the table.

(c) The string breaks. Describe what happens to the particle on the table. [A]

22 A car of mass 1.2 tonnes is travelling around a roundabout, at a steady speed of 12 m s^{-1}. The friction force that is acting on the car has a magnitude equal to 90% of the magnitude of the normal reaction on the car. Assume that the car can be modelled as a particle and that the road surface is horizontal.

(a) Draw a diagram to show the forces acting on the car if there is assumed to be no air resistance.

(b) Find the radius of the circle described by the car as it travels around the roundabout.

(c) The diagram shows the air resistance force that actually acts on the car as it moves on the roundabout, but that has been ignored in (a) and (b).

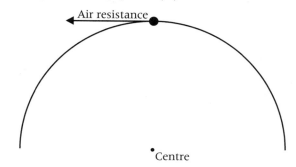

On a copy of the diagram draw a vector to show the resultant force on the car, and hence a vector to show the direction of the friction force on the car (i.e. the force between the tyres and the road). [A]

5.4 Further circular motion

In this section you will consider examples of circular motion in situations where the forces acting are not just horizontal and vertical. The first of these examples is a simple situation, known as the conical pendulum.

Conical pendulum

Consider a particle, of mass m kg, which is suspended from a fixed point A by a light, inextensible string of length l. If the particle moves in a horizontal circle, then the string describes the curved surface of a cone.

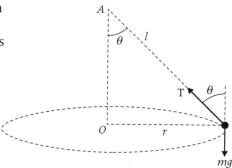

The centre of the circle O will be vertically below A. The radius of the circle will be $l \sin \theta$, where θ is the angle between the string and the vertical.

Using $F = ma$ radially: $T \sin \theta = mr\omega^2$
or $T \sin \theta = ml \sin \theta \, \omega^2$
   $T = ml\omega^2$

Resolving vertically: $T \cos \theta = mg$

From these two expressions it is possible to eliminate T to give
$$\cos \theta = \frac{mg}{ml\omega^2} = \frac{g}{l\omega^2}.$$
Note that, from this result, the angle does not depend on the mass. It also allows us to predict the angle for various combinations of l and ω.

Worked example 5.5

A ball, of mass 400 grams, is attached to one end of a light inelastic string of length 0.75 m. The other end of the string is fixed to a point A. The ball moves in a horizontal circle, at 120 rpm. Find the tension in the string and the angle between the string and the vertical.

Solution

The diagram shows the forces acting on the ball.
First convert the angular speed to radians per second:
$$\omega = \frac{120 \times 2\pi}{60} = 4\pi \text{ rad s}^{-1}$$
Using $F = ma$ radially $a = r\omega^2$ gives
$$T \sin \theta = 0.4 \times 0.75 \sin \theta \times (4\pi)^2$$
$$T = 47.37 \text{ N}$$
Resolving vertically
$$T \cos \theta = 0.4g$$
$$\cos \theta = \frac{0.4 \times 9.8}{47.37}$$
$$\theta = 85.3°$$

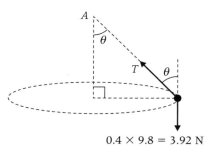

Note that the radius of the circle is $0.75 \sin \theta$.

Worked example 5.6

A particle P, of mass m, is connected to two light inextensible strings of equal length l. The other ends of the strings are attached to fixed points A and B, a distance l apart with A vertically above B. The particle P rotates in a horizontal circle with angular speed ω, and both strings taut. Find the tension in each string and show that
$$\omega^2 > \frac{2g}{l}$$

Solution

The diagram shows the two tensions that act, along with the weight of the particle.

Using $F = ma$ radially gives
$$T_1 \sin 60° + T_2 \sin 60° = ml \sin 60° \omega^2$$
$$T_1 + T_2 = ml\omega^2$$

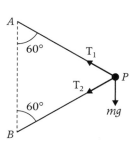

Resolving vertically gives

$$T_1 \cos 60° = T_2 \cos 60° + mg$$
$$T_1 - T_2 = 2mg$$

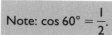
Note: $\cos 60° = \dfrac{1}{2}$.

You now have a pair of simultaneous equations which when added give

$$2T_1 = ml\omega^2 + 2mg$$
$$T_1 = mg + \tfrac{1}{2}ml\omega^2$$

Subtracting the equations gives

$$2T_2 = ml\omega^2 - 2mg$$
$$T_2 = \tfrac{1}{2}ml\omega^2 - mg$$

So the tensions in the strings are functions of m, g and ω. In particular the tension in the top string is positive for all values of ω, but this is not the case in the lower string. If the angular speed is not great enough the lower string becomes slack. Hence if both strings are taut

$$T_2 > 0$$
$$\tfrac{1}{2}ml\omega^2 - mg > 0$$
$$\omega^2 > 2\dfrac{g}{l}$$

Worked example 5.7

A sphere P, of mass m, moves in a horizontal circle, with angular velocity ω, on the inside, smooth surface of an inverted cone, of semi-vertical angle α as shown. Find the radius of the circle in terms of α, ω and g.

Solution

The diagram shows the two forces acting on the sphere, which is modelled as a particle. These forces are the weight and the normal reaction, which is perpendicular to the surface of the cone.

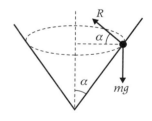

Resolving vertically gives

$$R \sin \alpha = mg$$

Using $F = ma$ radially gives

$$R \cos \alpha = mr\omega^2$$

These two expressions can then be divided as below

$$\dfrac{R \sin \alpha}{R \cos \alpha} = \dfrac{mg}{mr\omega^2}$$

$$\tan \alpha = \dfrac{g}{r\omega^2}$$

Now rearrange to make r the subject of this expression.

$$r = \dfrac{g}{\omega^2 \tan \alpha}$$

Worked example 5.8

An aircraft banks as it turns in a horizontal circle of radius 500 m. If the speed of the aircraft is 300 km h^{-1}, find the angle to the horizontal at which the aircraft must be banked.

Solution

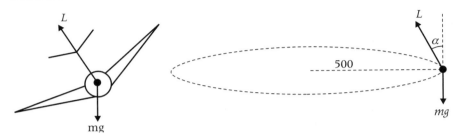

To model this situation assume that the lift force acts through the centre of gravity of the aircraft. The aircraft can thus be treated as a particle. The diagram shows the forces acting on the aeroplane.

Resolving vertically gives

$L \cos \alpha = mg$ [1]

Using $F = ma$ radially gives,

$L \sin \alpha = m \times \dfrac{v^2}{500}$,   [2]

where $v = 300$ km h$^{-1} = \dfrac{300 \times 1000}{3600} = 83.3$ m s^{-1}.

Dividing equation [2] by equation [1], and substituting for v gives

$\dfrac{\sin \alpha}{\cos \alpha} = \dfrac{v^2}{500g} = \dfrac{83.3^2}{500 \times 9.8} = 1.42$

$\tan \alpha = 1.42$
$\alpha = 55°$ to the nearest degree.

EXERCISE 5C

1 A particle, of mass 2 kg, is attached to a fixed point, A, by a light inextensible string of length l m, which is inclined at an angle θ to the vertical. The particle moves in a horizontal circle with angular speed ω. The tension in the string is T N.

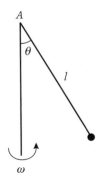

(a) If $l = 0.5$ m and $\theta = 20°$, find T and ω.

(b) If $l = 1$ m and $\omega = 5$ rad s^{-1}, find T and θ.

(c) If $\omega = 2$ rad s^{-1} and $\theta = 60°$, find l and T.

2 A particle, of mass 2 kg, is attached to a fixed point, A, by a light inextensible string of length 50 cm. The particle moves in a horizontal circle of radius 10 cm. Find the tension in the string and the particle's angular speed.

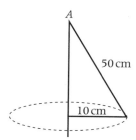

3 A particle, of mass 3 kg, is attached to a fixed point, A, by an inelastic string of length 70 cm. The particle moves in a horizontal circle with angular speed of 60 rpm. Find the tension in the string and the radius of the circle.

4 A particle, P, of mass 4 kg, is attached by a light inextensible string, of length 3 m, to a fixed point. The particle moves in a horizontal circle with an angular speed of 2 rad s^{-1}. Calculate:
  (a) the tension in the string,
  (b) the angle the string makes with the vertical,
  (c) the radius of the circle.

5 A particle, of mass 5 kg, is attached to a fixed point by a string of length 1 m. It describes horizontal circles of radius 0.5 m. Calculate the tension in the string and the speed of the particle.

6 A bead, of mass m, is threaded on a string of length 8 m, which is light and inextensible. The free ends of the string are attached to two fixed points separated vertically by a distance, which is half the length of the string, the lower fixed point being on a smooth, horizontal table. The bead is made to describe horizontal circles on the table around the lower fixed point, with the string taut.
  (a) Show that the radius of the circle is 3 m.
  (b) What is the maximum value of ω, the angular speed of the bead, if it is to remain in contact with the table?

7 A fairground ride consists of a platform, which rotates horizontally. Ropes hang from the platform and people sit in cradles, suspended by the ropes. One rope is 5 m long and hangs from a point 5 m from the centre of rotation. A child sits in the cradle, which rotates in a horizontal circle. If the angle between the rope and the vertical is $\tan^{-1}\left(\frac{3}{4}\right)$ find the angular speed of the ride.

8 A conical pendulum consists of a small sphere of mass 3 kg attached to the end of a light, inextensible string of length 0.6 m. The sphere moves in a horizontal circle at a constant speed. Model the sphere as a particle.

(a) The string is inclined at an angle of 30° to the vertical, as shown in the diagram.
  (i) Find the tension in the string.
  (ii) Show that the angular speed of the sphere is 4.34 rad s^{-1}, to three significant figures.

(b) The angular speed of the particle is now doubled. Find the new angle between the string and the vertical, giving your answer to the nearest degree. [A]

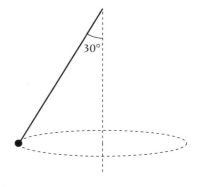

9 A particle, P, of mass 2 kg, is attached to two strings of lengths 0.2 m and 0.1 m. The strings are fixed to the particle and to the points A and B, respectively. The point A is directly above the point B. The particle describes a horizontal circle, centre B and radius 0.1 m, at a speed of 4 m s^{-1}. The particle and strings are shown in the diagram.

(a) Calculate the magnitude of the acceleration of the particle.
(b) Find the tension in the upper string.
(c) Find the tension in the lower string. [A]

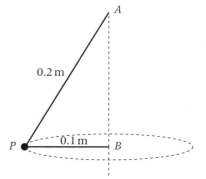

10 A particle, P, of mass 2 kg, is attached by a light, inextensible string, of length 0.5 m, to the point O. The particle moves in a circular path on a smooth horizontal table. The centre of the circle is vertically below O. The speed of the particle is 0.3 m s^{-1}, and the string is at an angle of 30° to the vertical.

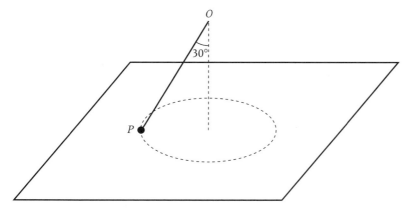

(a) Draw a diagram to show the forces acting on the particle.
(b) Show that the tension in the string is 1.44 N.
(c) Find the magnitude of the normal reaction force on the particle. [A]

11 One end of a light inextensible string of length l is fixed at a point A and a particle P of mass m is attached to the other end. The particle moves in a horizontal circle with constant angular speed ω. Given that the centre of the circle is vertically below A and that the string remains taut with AP inclined at an angle α to the downward vertical, find $\cos \alpha$ in terms of l, g and ω. [A]

12 A particle P is attached to one end of a light inextensible string of length 0.125 m, the other end of the string being attached to a fixed point, O. The particle describes, with constant speed, and with the string taut, a horizontal circle whose centre is vertically below O. Given that the particle describes exactly two complete revolutions per second find, in terms of g and π, the cosine of the angle between OP and the vertical. [A]

13 A particle, P, of mass m, moves in a horizontal circle, with uniform speed v, on the smooth inner surface of a fixed thin, hollow hemisphere of base radius a. The plane of motion of P is a distance $\dfrac{a}{4}$ below the horizontal plane, containing the rim of the hemisphere. Find, in terms of m, g and a, as appropriate, the speed v and the reaction of the surface on P.

A light inextensible string is now attached to P. The string passes through a small smooth hole at the lowest point of the hemisphere, and has a particle of mass m hanging at rest at the other end. Given that P now moves in a horizontal circle, on the inner surface of the hemisphere with uniform speed u, and that the plane of the motion is now distant $\dfrac{a}{2}$ below the horizontal plane of the rim, prove that:
$$u^2 = 3ga$$
[A]

14 The point A is vertically above point B, and a distance $5a$ from it. A particle, P, of mass m, is attached to A by a light inextensible string of length $4a$. The particle is also attached to B by a light inextensible string of length $3a$. P moves in a horizontal circle with both strings taut. Find the tension in the strings and show that:
$$\omega^2 \geq \frac{5g}{16a}$$
[A]

15 An aeroplane, of mass m kg, describes a horizontal circle of radius r m at a constant speed of v m s^{-1}.
  (a) Find the magnitude of the resultant force on the aeroplane if $m = 2000$, $v = 50$ and $r = 500$.

(b) A lift force of magnitude L N acts on the aeroplane. This force lies in the vertical plane that contains the aeroplane and the centre of the circle and acts at an angle α to the vertical when the aeroplane is flying in a circle at a constant speed. Find L and α in terms of v, r and g.

(c) Describe how L and α would change if the radius of the circle were reduced, but the speed of the aeroplane remained unchanged. [A]

16 A child, of mass 40 kg, swings on the end of a rope of length 4 m, moving in a horizontal circle at a constant speed. The rope is at angle of 30° to the vertical. The centre of the circle is at a height of 1.5 m above a point O on the ground. Model the child as a particle at the end of the rope.

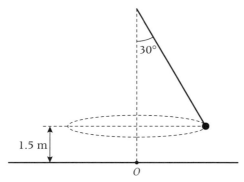

(a) Find:
　(i) the tension in the rope,
　(ii) the acceleration of the child,
　(iii) her speed.

The child lets go of the rope and falls freely under gravity.

(b) Find the time that it takes for the child to reach the ground and state a reason why, in reality, this is an overestimate.

(c) Find the maximum distance that the child could land from the point O. Draw a diagram to show the region in which the child could land and state the distance of the boundaries of this region from the point O. [A]

17 A particle, of mass m, is attached to one end of a light, inextensible string of length l. The other end of the string is fixed at P. The particle moves in a horizontal circle of radius r at a constant speed v, as shown in the diagram.

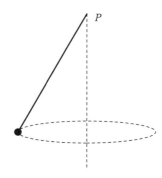

(a) Show that the tension in the string, T, is given by
$$T = \frac{mgl}{\sqrt{l^2 - r^2}}$$

(b) Find v^2 in terms of r, g and l.

(c) A second light, inextensible string, of length l, is then attached to the particle and to Q, a fixed point directly below P, as shown in the diagram. The particle moves in a horizontal circle of radius r at a constant speed V, with both strings taut.

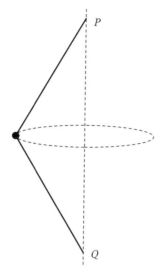

Find the tension in the upper string in terms of m, g, l, r and V. **[A]**

18 A fairground ride consists of a circular drum. The drum is a circular cylinder with vertical sides and of radius 2 m. It rotates about a vertical axis, and when it has reached its normal operating speed, the floor of the drum is removed so that the people on the ride move in a horizontal circle 'stuck' to the inside of the drum. At its normal operating speed, the drum completes 1 revolution every 2 seconds. Model a person on the ride as a particle.

(a) Find the acceleration of a person on the ride, at its normal operating speed.

(b) Draw and label a diagram to show the forces acting on a person on the ride, when the floor has been removed. Find the magnitude of the reaction and friction forces acting on a person of mass 70 kg in this situation.

(c) The operators propose a new design in which the sides of the drum are at angle θ to the horizontal as shown in the diagram.

Assume that people on the ride still move in a circle of radius 2 m and at the same angular speed as before. Show that the magnitude of the friction force, F, acting on a person of mass m, is given by
$$F = mg \sin \theta - 2\pi^2 m \cos \theta.$$
Hence show that F is zero when $\tan \theta = \dfrac{2\pi^2}{g}$. **[A]**

19 The diagram shows a truncated cone. The radius of the cone decreases from 40 cm to 30 cm and the height of the truncated cone is 50 cm.

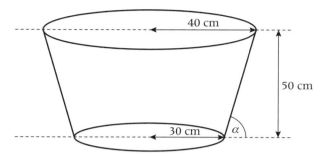

The angle between the sloping surface of the cone and the horizontal is α.

(a) Find $\tan \alpha$.

A coin is projected into the truncated cone, so that it rolls round on the inside surface.

When the coin is inside the truncated cone, it is modelled as a particle that **slides** on a smooth surface and it is assumed that there is no air resistance. The coin is assumed to move in a horizontal circle, on the inside surface of the cone, at a constant speed.

(b) Show that if the coin moves in a circle of radius r m, its speed, v m s^{-1}, is given by $v = \sqrt{rg \tan \alpha}$.

(c) Find the range of speeds for which the coin can be modelled as a particle describing a horizontal circle inside the truncated cone.

In reality the coin slows down, gradually moves down the cone, and eventually drops out of the bottom. Assume that the coin moves in a horizontal circle, of radius 40 cm, when it enters the truncated cone and that later it moves in a circle, of radius 30 cm, just before it leaves the cone. The mass of the coin is 10 grams.

(d) Find the total energy lost by the coin, as it moves from the top to the bottom of the truncated cone.

(e) State one aspect of the motion of the coin that has been ignored by the use of the particle model. [A]

Key point summary

1. $1 \text{ rpm} = \dfrac{2\pi}{60} \text{ rad s}^{-1}$. *p92*

2. The velocity has magnitude $r\omega$ and is directed along a tangent to the circle. *p92*

3. The acceleration of a particle moving with circular motion at a constant speed is directed towards the centre of the circle and has magnitude $r\omega^2$. It can be useful to express the magnitude of acceleration in terms of speed, v, rather than ω. From $v = r\omega$, $\omega = \dfrac{v}{r}$ and by substituting this, the expression for the magnitude of acceleration becomes
$$a = r\left(\dfrac{v}{r}\right)^2 = \dfrac{v^2}{r}.$$
p93

4. When a particle moves in a horizontal circle with constant speed, there are two principles that can always be applied: *p95*
   - the resultant of vertical components of forces must be zero;
   - $F = ma$ can be applied radially.

Test yourself

1. A coin, of mass 20 grams, is placed on a turntable that rotates at 50 rpm. The coefficient of friction between the coin and the turntable is 0.6. *Section 5.1*
   - (a) Find the maximum distance that the coin can be placed from the centre of the turntable if the coin is to travel in a circle.
   - (b) Find the magnitude of the friction force in this case.

2. A car travels round a roundabout on a circular path of radius 150 m. The mass of the car is 1400 kg and it travels at 15 m s^{-1}. *Section 5.3*
   - (a) Find the magnitude of the friction force on the car.
   - (b) Find an inequality that μ, the coefficient of friction between the tyres and the road, must satisfy.

3. A ball, of mass 300 grams, is attached to one end of a light string of length 60 cm. The ball describes a horizontal circle with the string at an angle of 20° to the vertical. *Section 5.2*
   - (a) Find the tension in the string.
   - (b) Find the speed of the ball.

Test yourself ANSWERS

1. (a) 21.4 cm; (b) 0.118 N.
2. (a) 2100 N; (b) $\mu \geqslant 0.153$.
3. (a) 3.13 N; (b) 0.856 m s^{-1}.

CHAPTER 6
Circular motion with variable speed

Learning objectives

After studying this chapter you should be able to:
- solve problems using the radial and tangential components of the acceleration
- use conservation of energy on problems where bodies move in vertical circles.

6.1 Introduction

Chapter 5 studied circular motion, where bodies moved with a constant speed. you saw that a body moving in a circle had an acceleration of magnitude $\dfrac{v^2}{r}$ or $r\omega^2$ directed towards the centre of the circle. In this chapter we will extend these ideas to bodies that move with variable speed. We will see that there is still a component of the acceleration towards the centre of the circle but that there is now also a component along the tangent.

6.2 Circular motion at variable speed

There is a component of acceleration directed towards the centre, which has magnitude $\dfrac{v^2}{r}$ or $r\omega^2$. In addition there is a component of acceleration along a tangent to the circle. The magnitude of this component is $\dfrac{dv}{dt}$. This additional component is present because the speed is changing.

Note that for constant speed, $\dfrac{dv}{dt} = 0$, and we simply have a single component towards the centre, which you used earlier in this book.

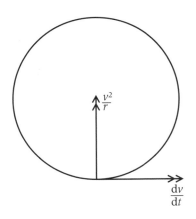

> The radial component of the acceleration is $\dfrac{v^2}{r}$ and the tangential component is $\dfrac{dv}{dt}$.

These two results are applied in the following examples.

Worked example 6.1

A car, of mass 1200 kg. is travelling on a roundabout of radius 50 m. At the point P, it is travelling at 15 m s^{-1} and its speed is increasing at 2 m s^{-2}.

Find the magnitude of the resultant force on the car:

(a) at the point P, (b) 5 s after it has left P.

Solution

(a) The car has a tangential component of acceleration of 2 m s^{-2}.

The radial component of the acceleration is given by
$$\frac{v^2}{r} = \frac{15^2}{50} = 4.5 \text{ m s}^{-2}.$$

The resultant of these two components will have magnitude
$$a = \sqrt{2^2 + 4.5^2} = 4.924 \text{ m s}^{-2}.$$

The resultant force will then have magnitude
$$F = 1200 \times 4.924 = 5910 \text{ N (to 3 sf)}.$$

(b) The tangential component of the acceleration of the car will still be 2 m s^{-2}.

The speed of the car will now be
$$v = 15 + 2 \times 5 = 25 \text{ m s}^{-1}.$$

The radial component of the acceleration can now be calculated.
$$\frac{v^2}{r} = \frac{25^2}{50} = 12.5 \text{ m s}^{-2}.$$

The magnitude of the resultant force on the car is then
$$F = 1200\sqrt{2^2 + 12.5^2} = 15\,200 \text{ N (to 3 sf)}.$$

Worked example 6.2

A car, of mass 1000 kg, is following a horizontal, circular path of radius 100 m. The coefficient of friction between the tyres of the car and the road is 0.7. Determine the maximum possible rate at which the speed of the car can increase, when it is travelling at 12 m s^{-1}.

Solution

We begin by assuming that the friction is the only horizontal force that acts on the car. The friction force provides both the radial and tangential components of the resultant force.

The maximum value of the friction force will be

$$F = \mu R$$
$$= 0.7 \times 1000 \times 9.8$$
$$= 6860 \text{ N}$$

The radial component of the resultant force will have magnitude

$$\frac{mv^2}{r} = \frac{1000 \times 12^2}{100}$$
$$= 1440 \text{ N}$$

As the magnitude of the resultant force is a maximum the tangential component of the resultant force will be given by

$$\sqrt{6860^2 - 1440^2} = 6710 \text{ N (to 3 sf)}.$$

The maximum rate of increase of speed is then given by

$$\frac{6710}{1000} = 6.71 \text{ m s}^{-2}.$$

EXERCISE 6A

1 A car travels on a horizontal circular track of radius 150 m. The car's speed increases at a constant rate of 0.5 m s^{-2}. Find the total acceleration of the car at the point where its speed is 15 m s^{-1}.

2 A car is travelling on a horizontal road which is an arc of a circle of radius 400 m. The car is travelling at a speed of 30 m s^{-1} when the brakes are applied so that the speed decreases at a rate of 1 m s^{-2}. Find the total acceleration of the car:

   (a) immediately after the brakes have been applied,

   (b) 5 s later.

3 A motorist is travelling on a circular track of radius 600 m and, when travelling at 35 m s^{-1} applies his brakes so that, 10 s later, the car is moving with speed 25 m s^{-1}.

   Assuming that the brakes produce a constant rate of decrease of speed, find the total acceleration of the car immediately after the brakes have been applied.

4 A car is travelling round a level road, which is an arc of a circle of radius 180 m. The speed of the car increases at the constant rate of 1.5 m s^{-2} and the magnitude of the total acceleration of the car at point A is 2.5 m s^{-2}. Find the speed of the car at this point.

5 A heavy lorry starts off on a curve of 250 m radius and its speed increases at a constant rate of 0.6 m s^{-2} from rest. Find the distance that the lorry will travel before its total acceleration reaches 0.8 m s^{-2}.

6. A particle, of mass m, can slide on a horizontal circular wire, of radius a, and the only force acting along the direction of motion is a force of constant magnitude $\dfrac{mV^2}{a}$ and opposing the motion.

   (a) The particle is projected along the wire with speed $5V$. Find the distance that it travels before coming to rest.

   (b) Find the initial acceleration of the particle.

7. A car, of mass 1200 kg, is driven at 10 m s^{-1} in a tight loop round a horizontal circle of radius 80 m. The tyres are limited to a total horizontal friction force of 10.6 kN. The driver then applies the brakes. What is the maximum possible deceleration?

6.3 Motion in a vertical circle

In this section we will consider bodies that move in vertical circles, where the increases in the energy of the bodies are due to gravity. For example, a person on a rope swing moves in a part of a vertical circle, as does a roller coaster that 'loops the loop'. In these types of example the results for the acceleration of the body used in the previous section can still be applied, but it will also be necessary to use conservation of energy to determine the speed of the body.

Worked example 6.3

A child, of mass 50 kg, holds one end of a rope of length 3 m. The other end of the rope is tied to a tree branch. Initially the child stands, at rest, on a bank with the rope horizontal. The child then swings on the rope and follows a circular path in a vertical plane.

(a) Find the tension in the rope when the rope is at angle θ to the vertical.

(b) Find the maximum tension in the rope and describe the position of the child when the tension is a maximum.

Solution

(a) The diagram shows the position of the child and the angle θ. The speed of the child can be determined using conservation of energy.

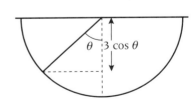

$$\text{Potential energy lost} = 50 \times 9.8 \times 3 \cos \theta = 1470 \cos \theta$$

$$\tfrac{1}{2} \times 50 v^2 = 1470 \cos \theta$$

$$v^2 = 58.8 \cos \theta$$

Now consider the forces acting on the child. These are shown in the diagram.

Resolving radially and applying Newton's second law gives

$$T - mg \cos \theta = \frac{mv^2}{r}.$$

Substituting the values for the mass and radius gives

$$T - 490 \cos \theta = \frac{50v^2}{3}.$$

The expression for v^2 obtained earlier can now be substituted.

$$T - 490 \cos \theta = \frac{50}{3} \times 58.8 \cos \theta$$

$$T = 1470 \cos \theta$$

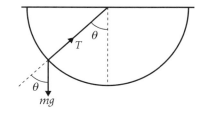

(b) As $\cos \theta$ has a maximum of 1 when $\theta = 0°$, the tension will have its maximum value when the position of the child corresponds to this value of θ. Referring back to the diagram it can be seen that $\theta = 0°$ when the rope is vertical and the child is at the lowest position. In this position the tension in the rope is 1470 N.

Worked example 6.4

A particle, of mass 5 kg, is placed at the point A at the top of a hemisphere, of radius 2 m and centre O. The hemisphere is fixed to a horizontal surface, as shown in the diagram. The particle is set into motion with an initial horizontal speed of 3 m s^{-1}. The particle leaves the surface of the hemisphere at the point B. Find the angle AOB.

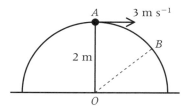

Solution

The diagram shows the forces acting on the particle and the angle θ.

Using conservation of energy the speed of the particle at the point shown can be determined.

Potential energy lost $= 5 \times 9.8 \times (2 - 2 \cos \theta) = 98(1 - \cos \theta)$

As this will be equal to the gain in kinetic energy, we have

$$98(1 - \cos \theta) = \frac{1}{2} \times 5v^2 - \frac{1}{2} \times 5 \times 3^2$$

$$v^2 = 39.2(1 - \cos \theta) + 9$$

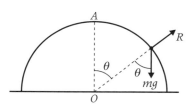

Resolving radially at the point shown in the diagram and applying Newton's second law gives

$$5g \cos \theta - R = \frac{5v^2}{2}$$

$$R = 5g \cos \theta - \frac{5v^2}{2}$$

The particle will leave the surface when $R = 0$. This can be expressed in terms of the speed of the particle as

$$0 = 5g \cos \theta - \frac{5v^2}{2}$$

$$v^2 = 2g \cos \theta$$

Now this result can be substituted into the expression for the speed obtained by considering energy, to give the required angle.

$$19.6 \cos \theta = 39.2(1 - \cos \theta) + 9$$

$$\cos \theta = \frac{39.2 + 9}{19.6 + 39.2}$$

$$\theta = 34.9°$$

Worked example 6.5

A toy car is released from rest on the loop-the-loop track shown in the diagram. The loop has a diameter of 50 cm. Determine the minimum height, h, from which the car must be released if the car is to loop the loop.

Solution

For the car to loop the loop it must remain in contact with the track at all times. The diagram shows the forces acting on the car.

Resolving radially and applying Newton's second law gives

$$R - mg \cos \theta = \frac{mv^2}{r}.$$

If the car stays in contact with the track at any point

$$R \geqslant 0$$

$$\frac{mv^2}{r} + mg \cos \theta \geqslant 0$$

$$v^2 \geqslant -rg \cos \theta$$

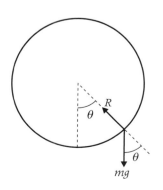

Note that cos θ varies between -1 and 1, so that taking $\theta = 180°$, for which cos $\theta = -1$, gives

$v^2 \geq rg$.

This means that the minimum speed of the car at the top of the loop is $\sqrt{rg}$ or $\sqrt{0.25 \times 9.8} = \sqrt{2.45}$.

Now conservation of energy can be used to find the initial height of the car, as the total energy at the top of the loop must be equal to the total initial energy.

$$mgh = \frac{1}{2}mv^2 + mg \times 2r$$

Using the values for v obtained above and the radius of the circle gives

$$9.8h = \frac{1}{2} \times 2.45 + 9.8 \times 0.5$$

$$h = 0.625 \text{ m}$$

When attempting problems involving vertical circular motion:

- use conservation of energy to find the speed of the object in the required position,
- resolve radially and apply Newton's second law.

EXERCISE 6B

1. A soldier, of mass 75 kg, swings on a light inextensible rope of length 6 m. The soldier is initially at rest and the angle between the rope and the vertical is 30°.

   (a) Find the maximum speed of the soldier and the tension in rope at that time.

   (b) Find the tension in the rope when the rope is at an angle of 20° to the vertical.

2. A bead is threaded onto a smooth vertical hoop of radius 2 m. The mass of the bead is 0.05 kg.

   (a) If the speed of the bead is 10 m s^{-1} at the lowest point of the hoop, find the speed of the bead at the top of the hoop and the force that the hoop exerts on the bead in this position.

   (b) Determine the force exerted on the bead by the hoop at the lowest point of the hoop.

3 An eskimo, of mass 60 kg, sits on the top of his igloo. The igloos is assumed to be a smooth hemisphere of radius 3 m. The eskimo is pushed so that he initially moves horizontally at 0.5 m s^{-1}. The diagram shows the angle θ when the eskimo is at the point P.

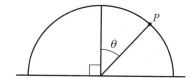

(a) Find the reaction force exerted on the eskimo when $\theta = 10°$,

(b) Show that the eskimo leaves the surface of the igloo and find the value of θ when this happens.

4 A child, of mass 35 kg, sits on a swing and swings freely through an angle of 30° on either side of the vertical. The ropes of the swing are 2.5 m long. Modelling the motion as that of a particle of mass 35 kg attached to an inextensible rope of length 2.5 m, find the speed of the child when the rope is vertical and also the tension in the rope at that instant.

5 An aeroplane is flown at a constant speed of 175 m s^{-1} in a vertical circle of radius 1000 m. Find the force exerted by the seat on the pilot, of mass 80 kg, at the lowest and highest points.

6 A man swings a bucket full of water in a vertical plane in a circle of radius 0.5 m. What is the smallest velocity that the bucket should have at the top of the circle if no water is to be spilt?

7 The diagram shows the rotating drum of a spin-dryer. The radius of the drum in 0.3 m. Find the angular velocity of the drum so that a small article of clothing drops off the drum when $\theta = 40°$. You may assume that the surface of the drum is such as to prevent slipping before loss of contact.

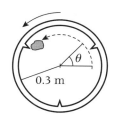

8 A bead, of mass m, is threaded on a smooth circular loop of wire of radius a and which is fixed in a vertical plane. The bead is released from rest at the end of a horizontal diameter. Find the reaction of the wire when the bead has turned through an angle θ.

9 A particle is released from rest at a point on the outer surface of a smooth sphere of radius a; the point of release is at height $\dfrac{a}{2}$ above the centre. Find the height above the centre at which the particle leaves the sphere.

10 On a child's toy, a small car is fired along a smooth track and loops the loop inside a section of track fixed in the form of a vertical circle of radius a and centre O, as shown in the diagram.

The car is travelling at speed u as it enters the circle at the lowest point. Modelling the car as a particle P, of mass m, find the reaction of the track on the particle in terms of m, g, a, θ and u when OP makes an angle θ with the downward vertical.

Deduce that the car will complete the vertical circle if $u > \sqrt{5ga}$. [A]

11 A bead, of mass m, is threaded onto a smooth circular ring, of radius r, which is fixed in a vertical plane. The bead is moving on the wire so that its speed, v, at the highest point of its path is half of its speed at the lowest point.

(a) Find v in terms of r and g.

(b) Find the reaction of the wire on the bead, in terms of m and g, when the bead is
   (i) at the highest point,
   (ii) $\frac{1}{2}r$ above its lowest point. [A]

12 A bead, of mass m, is threaded onto a smooth circular ring, of radius r, which is fixed in a vertical plane. The bead is moving on the wire so that its speed at the lowest point of its path is four times its speed, v, at the highest point.

(a) Find v in terms of r and g.

(b) Find the reaction of the wire on the bead when the bead is at its lowest point. [A]

13 In crazy golf, a golf ball is fired along a smooth track and loops the loop inside a section of track.

Model this loop as a vertical circle of radius a and centre Q, as shown in the diagram.

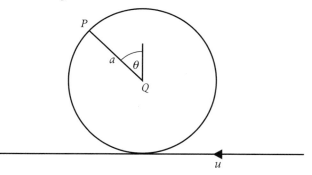

The golf ball is travelling at speed u as it enters the circle at the lowest point.

Model the ball as a particle P, of mass m.

(a) Show that the reaction of the track on the particle when QP makes an angle of θ with the upward vertical is

$$\frac{mu^2}{a} - 3mg\cos\theta - 2mg.$$

(b) Given that the ball completes a vertical circle inside the track, show that $u \geq \sqrt{(5ag)}$ [A]

14 A smooth hemisphere of radius l and centre Q, lies with its plane face fixed to a horizontal surface. A particle, P, of mass m, can move freely on the surface of the hemisphere.

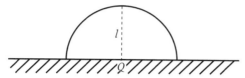

The particle is set in motion along the surface of the hemisphere with a speed u at the highest point of the hemisphere.

(a) Show that, while the particle is in contact with the hemisphere, the velocity of the particle when PQ makes an angle θ to the vertical, is

$$(u^2 + 2gl[1 - \cos\theta])^{\frac{1}{2}}.$$

(b) Find, in terms of l, u and g, the cosine of the angle θ when the particle leaves the surface of the hemisphere. [A]

15 A bead, of mass m, is threaded onto a smooth circular ring, of radius r, which is fixed in a vertical plane. The bead is moving on the wire. Its speed, v, at the highest point of its path is one quarter of its speed at the lowest point.

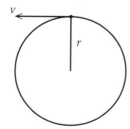

(a) Show that $v = \sqrt{\dfrac{4}{15}gr}$.

(b) Find the reaction of the wire on the bead, in terms of m and g, when the bead is:
   (i) at the highest point,
   (ii) $\frac{1}{2}r$ below its highest point. [A]

16

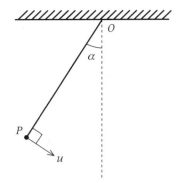

The diagram shows a particle, *P*, of mass *m*, which is attached by a light inextensible string, of length *l*, to a fixed support *O*. The particle moves in a vertical plane. It is initially set in motion with speed *u* at right angles to *OP*, from the position where *OP* makes an angle α with the downward vertical through *O*.

(a) Show that when *OP* makes an angle θ with the downward vertical through *O*, the speed, *v*, of the particle is given by
$$v^2 = u^2 + 2gl(\cos\theta - \cos\alpha).$$

(b) At an adventure playground, a girl, of mass 40 kg, swings on the end of a rope of length 5 m. The motion is in a vertical plane. Initially the rope makes an angle of 30° with the downward vertical and the girl has a speed of 2 m s^{-1} at right angles to the rope.
  (i) Show that the maximum speed of the girl during the motion is approximately 4.1 m s^{-1}.
  (ii) Determine the maximum angle that the rope makes with the downward vertical.
  (iii) Find the maximum tension in the rope.
  (iv) State one modelling assumption which you have used in this problem. [A]

17

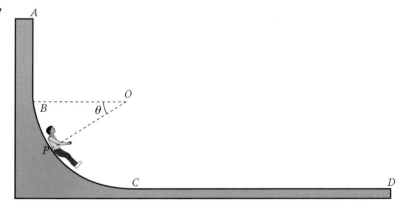

The diagram shows a vertical cross section of a new adventure slide at a theme park. It consists of three sections *AB*, *BC* and *CD*.

Section *AB* is smooth and vertical and has length r.

Section *BC* is smooth and forms a quarter of a circle. This circle has centre *O* and radius r. The radius *OB* is horizontal and *OC* is vertical.

Section *CD* is rough, straight and horizontal. It is of length $4r$.

Steve, who has mass m, starts from rest at *A* and reaches speed u at the point *B*. He remains in contact with the surface until he reaches *D*.

It can be assumed that Steve can be modelled as a particle throughout the motion.

(a) Find u^2 in terms of g and r.

(b) Steve reaches the point *P*, between *B* and *C*, where angle $POB = \theta$, as shown in the diagram. His speed at *P* is v.
  (i) Show that $v^2 = 2gr(1 + \sin\theta)$.
  (ii) Draw a diagram showing the forces acting on Steve when he is at the point *P*.
  (iii) Find an expression for the normal reaction, R, on Steve when he is at the point *P*. Give your answer in terms of m, g and θ.

(c) Show that, as Steve crosses *C*, there is a reduction in the normal reaction of magnitude $4mg$.

(d) Between *C* and *D*, Steve decelerates uniformly and comes to rest at the point *D*. Find his retardation. [A]

18 Maria is modelling the motion of a toy car along a loop-the-loop track.

The track, *ABCD*, can be modelled as a continuous smooth surface contained in a vertical plane. The highest point on the track is *A*, which is a distance h above the floor. The loop is a circle of radius r with *BC* as a diameter, and with *C* vertically above *B*. The track ends at the point *D*, as shown in the diagram.

Maria is trying to determine a connection between h and r in the case where the car stays only just in contact with the track at *C*.

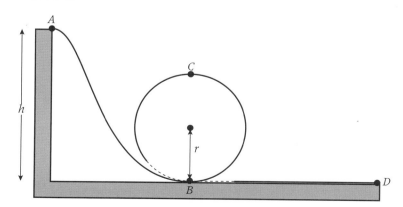

Maria models the car as a particle of mass m, which starts from rest at A. In the case where the car stays only just in contact with the track at C, it has speed u at B and speed v at C.

(a) By considering the forces on the car at C, show that $v^2 = rg$.

(b) Find an expression for u^2 in terms of g and r.

(c) Show that $h = kr$, where k is a constant to be determined.

(d) Suggest one improvement that could be made to the model in order to refine the solution. [A]

19

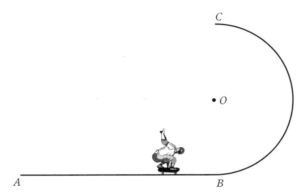

The diagram shows part of a track used by a skateboarder. The track consists of a horizontal part AB and a curved part BC. The curved part can be modelled as a smooth semicircular arc with centre O and radius r, with C vertically above B.

The skateboarder may be modelled as a particle moving on the track. He has a speed of $\sqrt{\dfrac{7gr}{2}}$ as he reaches the point B.

(a) The point D is on the arc BC and OD makes an angle of $60°$ with the upward vertical. Find the speed of the skateboarder, in terms of g and r, as he reaches D.

(b) Show that, at the point D, the skateboarder is about to lose contact with the track. [A]

20 James and Emma are carrying out separate experiments, using two identical particles. James attaches his particle to a light inelastic string of length l.

Emma threads her particle onto a smooth ring, of radius l, which is fixed in a vertical plane. James and Emma then measure the minimum speed, at the lowest point, necessary for the particle to be able to make complete circles in a vertical plane.

(a) What should James find to be the minimum speed?

(b) What should be the difference between the two speeds which they find? [A]

Key point summary

1. The radial component of the acceleration is $\dfrac{v^2}{r}$ and the tangential component is $\dfrac{dv}{dt}$.   p112

2. When attempting problems involving vertical circular motion: p118
   - use conservation of energy to find the speed of the object in the required position,
   - resolve radially and apply Newton's second law.

Test yourself

	What to review
1 A car, of mass 1200 kg, follows a circular path of radius 120 m. The speed of the car is increasing at 0.5 m s^{-2}. Find the magnitude of the resultant force on the car when it is travelling at 10 m s^{-1}.	Section 6.2
2 A bead, P, of mass m, is threaded onto a smooth circular hoop of radius 0.75 m and centre O. The bead is set into motion with a speed of 3 m s^{-1} at the top of the hoop.	Section 6.3

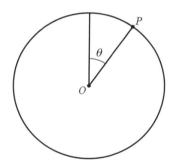

(a) Find the speed of the bead when it is in the position shown in the diagram.

(b) Find the position of the bead when the reaction force exerted on the bead by the hoop is equal to half of the reaction force exerted by the hoop on the bead at its lowest point.

Test yourself ANSWERS

1 1166 N.

2 (a) $v = \sqrt{9 + 14.7(1 - \cos\theta)}$;

(b) 87.9°.

Application of differential equations in mechanics

Learning objectives

After studying this chapter you should be able to:

- use $\dfrac{dv}{dt}$ to form and solve differential equations to obtain relationships between velocity and time.

In your earlier work you will have encountered situations where the acceleration of a body is given as a function of t.

In these cases the function can be integrated directly to find the velocity and again to find the displacement.

For example, if you have $\dfrac{dv}{dt} = f(t)$, then v can be found using integration as $v = \int f(t)\, dt$.

However there are other cases that are not so straightforward, where the acceleration might depend on the velocity of the body. For example the acceleration of a skydiver varies with his or her velocity. These types of problems give rise to differential equations of the form $\dfrac{dv}{dt} = f(v)$. Differential equations of this type can often be solved using the technique of separation of variables. In this case you would obtain

$$\frac{1}{f(v)} \times \frac{dv}{dt} = 1$$

$$\int \frac{1}{f(v)}\, dv = \int 1\, dt$$

$$= t + c$$

Problems of this type are very common and some are considered in the following examples.

Worked example 7.1

A boat of mass 500 kg is travelling at 8 m s^{-1}, when its outboard motor falls off. The boat travels in a straight line until it stops. Assume that it experiences a resistive force of magnitude $25v$ N when travelling at v m s^{-1}.

(a) Find an expression for the velocity of the boat at time t seconds after the motor has fallen off, in terms of t, the time since the motor fell off.

(b) How far does the boat travel in the first 5 seconds after the motor falls off?

Solution

(a) There will be a single horizontal force acting on the boat, which will be the resistance force, as shown in the diagram. As this opposes the motion, the resultant force can be written as $-25v$. Then applying Newton's second law gives

$$500 \frac{dv}{dt} = -25v$$

or

$$\frac{1}{v} \times \frac{dv}{dt} = -\frac{1}{20}$$

An integral can now be formed and evaluated

$$\int \frac{1}{v} dv = \int -\frac{1}{20} dt$$

$$\ln|v| = -\frac{t}{20} + c$$

Since v will always be positive in this problem we can discard the modulus signs and solve the equation for v, to give

$$\ln v = -\frac{t}{20} + c$$

$$v = e^{-\frac{t}{20} + c} = e^{-\frac{t}{20}} = e^c$$

$$= Ae^{-\frac{t}{20}}$$

Note how A has been used to replace e^c and simplify the equation. Initially the boat was moving at 8 m s^{-1}, so substituting $v = 8$ and $t = 0$, will allow us to find A.

$$8 = Ae^0$$

$$A = 8$$

$$v = 8e^{-\frac{t}{20}}$$

(b) The distance travelled can be found by integrating the expression for the velocity, with limits of integration of 0 and 5.

$$\text{Distance} = \int_0^5 8e^{-\frac{t}{20}}\, dt$$

$$= \left[-160e^{-\frac{t}{20}} \right]_0^5$$

$$= (-160e^{-\frac{5}{20}}) - (-160e^0)$$

$$= 35.4 \text{ m (to 3 sf)}$$

Worked example 7.2

A particle, of mass 6 kg, moves on a smooth horizontal wire through a fluid. The fluid exerts a resistance force of magnitude $7.5\, v^2$ N on the particle when it is moving with velocity v m s^{-1}. The initial velocity of the particle is 50 m s^{-1}.

(a) Find an expression for v at time t seconds.

(b) Find the time when the velocity of particle is 10 m s^{-1}.

Solution

(a) As the resultant force will simply be $-7.5\,v^2$, Newton's second law can be applied to give

$$6\frac{dv}{dt} = -7.5\,v^2$$

$$\frac{dv}{dt} = -\frac{7.5\,v^2}{6} = -\frac{5}{4}v^2$$

Separating the variables leads to

$$\int \frac{1}{v^2}\,dv = \int -\frac{5}{4}\,dt$$

Then integrating, and including a constant of integration gives

$$-\frac{1}{v} = -\frac{5}{4}t + c$$

or

$$\frac{1}{v} = \frac{5}{4}t - c.$$

The value of c can now be determined using $t = 0$ and $v = 50$.

$$\frac{1}{50} = -c$$

$$c = -\frac{1}{50}$$

Hence,
$$\frac{1}{v} = \frac{5}{4}t + \frac{1}{50}.$$

Now v can be made the subject of the expression

$$\frac{1}{v} = \frac{125t}{100} + \frac{2}{100}$$

$$\frac{1}{v} = \frac{125t + 2}{100}$$

$$v = \frac{100}{125t + 2}$$

(b) You need to solve the equation

$$10 = \frac{100}{125t + 2}$$

or

$$1 = \frac{10}{125t + 2}$$

$$125t + 2 = 10$$

$$125t = 8$$

$$t = \frac{8}{125} = 0.064 \text{ seconds}.$$

Worked example 7.3

A particle, of mass m kg, experiences a resistance force of magnitude mkv^3 N when it is moving with speed v and where k is a constant.

Find an expression for the velocity of the particle at time t seconds, if the velocity of the particle is U when $t = 0$.

Solution

The resultant force on the particle will be $-mkv^3$, and using Newton's second law this gives

$$m\frac{dv}{dt} = -mkv^3$$

$$\frac{dv}{dt} = -kv^3$$

Separating the variables and integrating gives

$$\int \frac{1}{v^3} dv = \int -k \, dt$$

$$-\frac{1}{2v^2} = -kt + c$$

$$\frac{1}{2v^2} = kt - c$$

The value of c can be determined by using $v = U$ when $t = 0$.

$$\frac{1}{2U^2} = -c, \text{ so that } c = -\frac{1}{2U^2}$$

Hence,

$$\frac{1}{2v^2} = kt + \frac{1}{2U^2}$$

Now this can be solved for v:

$$\frac{1}{v^2} = \frac{2ktU^2}{U^2} + \frac{1}{U^2}$$

$$\frac{1}{v^2} = \frac{2ktU^2 + 1}{U^2}$$

$$v^2 = \frac{U^2}{2ktU^2 + 1}$$

$$v = \sqrt{\frac{U^2}{2ktU^2 + 1}}$$

EXERCISE 7A

1. A particle, of mass 2 kg, is set into motion on a smooth horizontal surface. Its initial velocity is 20 m s^{-1}. When moving at v m s^{-1}, it experiences a resistance force of magnitude $5v$ N.

   (a) Find an expression for the velocity of the particle after t seconds.

   (b) Find an expression for the distance travelled by the particle after t seconds.

   (c) How far does the particle travel?

2. A particle, of mass 4 kg, moves horizontally along a straight line and is subject to a resistance force of magnitude $2v^2$ N, when moving at v m s^{-1}. Find an expression for the velocity of the particle at time t seconds, if the initial velocity of the particle is

   (a) 5 m s^{-1}, (b) 20 m s^{-1}.

3. A torpedo is fired horizontally from a stationary submarine at a speed of 80 m s^{-1}. The torpedo has a mass of 100 kg and is subject to a resistance force of magnitude $50v$ N, when travelling at a speed of v m s^{-1}.

   (a) Find an expression for the speed of the torpedo at time t seconds.

   (b) Find an expression for the distance travelled by the torpedo in t seconds.

   (c) Find the distance travelled by the torpedo when its speed has dropped to 20 m s^{-1}.

4 A particle of mass 4 kg, slides on a smooth surface subject to a resistance force of magnitude $10v$ N, when travelling at v m s^{-1}. The particle has an initial speed of 20 m s^{-1}.

   (a) Find an expression for the velocity of the particle when it has been travelling for t seconds.

   (b) How far does the particle travel in the first 2 seconds of its motion?

5 A boat, of mass 400 kg, is initially travelling at 6 m s^{-1}, when its engine stops. The boat then slows down moving in a straight line. If the magnitude of the resistance force is $2v^2$ N, when the speed of the boat is v m s^{-1}, find how long it takes for the boat's speed to drop to 1 m s^{-1}.

6 A bullet, of mass 50 grams, is fired horizontally at an initial speed of 120 m s^{-1}. It hits a target 0.4 seconds after it was fired. The bullet is subject to a resistance force of magnitude $0.001v^2$ N. Assume that any vertical motion of the bullet can be neglected. Find the speed of the bullet when it hits the target.

7 A particle experiences a retardation of $0.005v^4$ m s^{-2}, when it travels at a speed of v m s^{-1}. It has an initial speed of 10 m s^{-1}.

   (a) Find the time when the speed has dropped to 5 m s^{-1}.

   (b) Find the speed after 2 seconds.

8 A model for the resistance force on a freewheeling car is based on the assumption that his force is proportional to the square of the speed of the car. An experiment is conducted in which a car is driven to a speed of 30 m s^{-1}, and then allowed to freewheel along a straight horizontal road until it stops. After the car has been freewheeling for 20 seconds, its speed is reduced to 15 m s^{-1}.

   (a) Find an expression for the velocity of the car when it has been freewheeling for t seconds.

   (b) Find the time it takes for the speed to drop to 5 m s^{-1}.

9 A particle of mass 20 kg, moves horizontally on a smooth surface subject to a resistance force of magnitude $4v$ N. It has an initial velocity of 20 m s^{-1}, when time $t = 0$ seconds.

   (a) Find the velocity of the particle when it has been travelling for t seconds.

   (b) Find the distance that the particle has travelled at time t.

10 A particle, of mass m, moves in a straight line on a smooth horizontal surface. As it moves it experiences a resistance force of magnitude kv^2, where k is a constant and v is the speed of the particle, at time t. The particle moves with speed U at time $t = 0$.

Show that $v = \dfrac{mU}{Ukt + m}$. [A]

11 A particle of mass m is moving along a straight horizontal line. At time t the particle has speed v. Initially the particle is at the origin and has speed U. As it moves the particle is subject to a resistance force of magnitude mkv^3.

(a) Show that $v^2 = \dfrac{U^2}{2kU^2 t + 1}$.

(b) What happens to v as t increases? [A]

12 A sphere, of mass m kg, moves in a fluid. It experiences a resistive force of magnitude $4mv$ N, when travelling at v m s^{-1}. Assume no other forces act on the sphere. The initial speed of the sphere is U.

(a) Find an expression for the velocity of the sphere after it has been moving for t seconds.

(b) Find an expression for the distance moved after t seconds.

13 A boat of mass 200 kg experiences a resistive force that has magnitude $25v$ N, when the boat is travelling at v m s^{-1}. Find expressions for the velocity and displacement of the boat at time t seconds, if the boat starts at the origin and has initial speed 10 m s^{-1}.

14 As a particle, of mass m, moves horizontally it experiences a resistance force of magnitude mkv^n, where v is the speed of the particle and k and n are positive constants with $n > 1$. The velocity of the particle was U at time $t = 0$ seconds. Show that

$$v^{n-1} = \dfrac{U^{n-1}}{1 + k(n-1)U^{n-1}t}.$$

Key point summary

1. $\dfrac{dv}{dt}$ can be used to form and solve differential equations to obtain relationships between velocity and time.   p126

Test yourself

	What to review
1 A particle, of mass 2 kg, slides horizontally on a smooth surface subject to a resistance force of magnitude $2v$ N, when travelling at v m s^{-1}. The initial speed of the particle is 10 m s^{-1}. (a) Find the speed of the particle after 2 seconds. (b) Find the distance travelled by the particle in the first 2 seconds	Section 7.1
2 A cyclist starts to freewheel, while travelling at 8 m s^{-1} on a straight horizontal road. The cyclist experiences a resistance force, of magnitude $3v^2$ N when travelling at a speed of v m s^{-1}. The cyclist and her cycle have a total mass of 75 kg. For how long has the cyclist been travelling by the time that her speed drops to 4 m s^{-1}?	Sections 7.2

Test yourself ANSWERS

1 (a) 1.35 m s^{-1}; **(b)** 8.65 m.

2 $3\tfrac{1}{8}$ s.

Exam style practice papers

MM2A

Time allowed 1 hour 15 minutes

Answer **all** questions

1 The diagram shows a beam of mass 20 kg and length 6 m. The beam rests on two supports as shown.

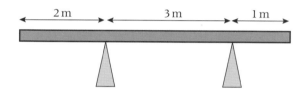

(a) Draw a diagram to show the forces acting on the beam. *(1 mark)*

(b) Show that the magnitude of the reaction force exerted by the right-hand support is 65.3 N. *(3 marks)*

(c) Find the magnitude of the reaction force exerted by the left-hand support. *(2 marks)*

2 At time t seconds the position vector of a particle is **r** metres, where

$$\mathbf{r} = 2\sin t\,\mathbf{i} + 3\cos t\,\mathbf{j}$$

The unit vectors **i** and **j** are perpendicular.

(a) (i) Find the velocity of the particle at time t seconds. *(2 marks)*

  (ii) Find the speed of the particle when $t = 2$. *(2 marks)*

(b) (i) Show that the magnitude of the acceleration is $\sqrt{4 + 5\cos^2 t}$. *(4 marks)*

  (ii) Hence state minimum magnitude of the acceleration. *(1 mark)*

3 An elastic string has natural length 4 m and modulus of elasticity 200 N.

  (a) Calculate the elastic potential energy of the string when it is stretched to a length of 8 m. (3 marks)

  (b) One end of the string is attached to a fixed point, P, at a height of 8 m above ground level. An object of mass 10 kg is attached to the string and released from rest at ground level directly below P.

   (i) Calculate the gain in gravitational potential energy as the object rises to a height of 4 m. (2 marks)

   (ii) Find the speed of the object when it is at a height of 4 m. (3 marks)

   (iii) Find the maximum height of the object above the ground. (3 marks)

4 A conical pendulum is made from a light, inextensible string of length 60 cm and a small sphere of mass 50 grams. The sphere is set into motion, so that it describes a horizontal circle with the string at an angle of 60° to the horizontal.

  (a) Show that the tension in the string is 0.566 N (correct to 3 sf). (2 marks)

  (b) Find the speed of the sphere. (5 marks)

5 The diagram shows a uniform lamina.

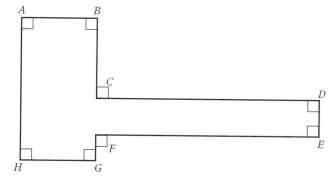

The dimensions of the lamina are:
  $AB = 3$ cm $EF = 11$ cm
  $BC = 3$ cm $FG = 1$ cm
  $CD = 11$ cm $GH = 3$ cm
  $DE = 2$ cm $AH = 6$ cm

  (a) Show that the centre of mass of the lamina is 5.35 cm from AH. (3 marks)

  (b) Find the distance of the centre of mass of the lamina from AB. (4 marks)

  (c) The lamina is suspended in equilibrium from the corner A. Find the angle between AB and the vertical. (3 marks)

6 A particle, of mass 3 kg, moves on the bottom surface of a tank. The particle moves on a straight, horizontal line. At time t seconds it is moving at v m s^{-1} and experiences a resistance force of magnitude $1.5v^2$ N.

(a) Show that $\dfrac{dv}{dt} = -\dfrac{v^2}{2}$. (2 marks)

(b) At time $t = 0$, the velocity of the particle is 7 m s^{-1}. Find an expression for the velocity of the particle at time t. (5 marks)

(c) Find the time when the velocity of the particle is 2 m s^{-1}, giving your answer as a fraction. (3 marks)

7 A toy moves on a loop-the-loop track as shown in the diagram.

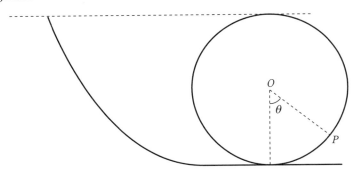

The loop is circular and is fixed in a vertical plane. The radius of the loop is 20 cm and the centre of the circle is at O. The car is released from rest at the same height as the top of the loop. At time t the car is at the point P on the loop. The angle between OP and the vertical is θ.

(a) Show that, while the car is on the loop, its speed v m s^{-1} is given by $v = \sqrt{3.92(1 + \cos\theta)}$. (3 marks)

(b) Find the value of θ at the point at which the car loses contact with the track. (5 marks)

MM2B

Time allowed 1 hour 30 minutes

Answer **all** questions

---

1. A ball of mass 0.4 kg is thrown vertically upwards at a speed of 6 m s^{-1}. The ball is at a height of 5 m above ground level, when it is thrown.

    (a) Calculate the initial kinetic energy of the ball. (2 marks)

    (b) Show that the kinetic energy of the ball when it hits the ground is 26.8 J and use this to find the speed of the ball when it hits the ground. (4 marks)

    (c) Use an energy method to find the maximum height of the ball. (3 marks)

    (d) State the assumption that you have had to make to obtain your answers to **(b)** and **(c)** of this question. (1 mark)

2. A particle moves so that, at time t seconds, its position vector is **r** metres, where

    $$\mathbf{r} = (7t^3 - 6)\mathbf{i} + (8t^2 - 5t + 9)\mathbf{j}.$$

    The unit vectors **i** and **j** are perpendicular.

    (a) (i) Find an expression for the velocity of the particle at time t. (2 marks)

    (ii) Find the time when the particle is travelling parallel to the unit vector **i**. (3 marks)

    (b) (i) Find an expression for the acceleration of the particle at time t. (2 marks)

    (ii) State the time when the magnitude of the acceleration is a minimum. (1 mark)

3. A car, of mass 1200 kg, travels round a roundabout of radius 50 m at a constant speed. The coefficient of friction between the tyres and the road surface is 0.7.

    (a) Find the maximum value of the friction force between the tyres and the road. (2 marks)

    (b) Find the maximum speed at which the car can travel round the roundabout. (4 marks)

4 The diagram shows a rectangular framework made up of four light rods, with particles attached at each corner. The diagram shows the framework and the mass of each of the particles.

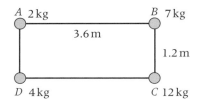

(a) Show that the centre of mass of the framework is a distance 0.768 m from *AB*. (3 marks)

(b) Find the distance of the centre of mass from *AD*. (4 marks)

(c) The framework is suspended from the corner *A* and is at rest in equilibrium. Find the angle between *AB* and the vertical. (3 marks)

(d) In reality the rods are not light. Would this change your answer to **(c)**? Explain why. (2 marks)

5 A lorry has mass 9 tonnes. As the lorry moves it experiences a resistance force of magnitude $60v$ newtons, where v is the speed of the lorry. The maximum speed of the lorry on a horizontal road is 40 m s^{-1}.

(a) Calculate the maximum power of the lorry. (3 marks)

(b) When the lorry is travelling at 10 m s^{-1} on a straight horizontal road it begins to accelerate. Find the maximum acceleration of the lorry at this speed on this road. (4 marks)

(c) The lorry is at rest at the bottom of a slope that is inclined at 6° to the horizontal. Find the maximum speed that the lorry could reach on this slope. (6 marks)

6 A ladder of length 4 m and mass 12 kg is in equilibrium, with its top against a smooth vertical wall and its base on rough horizontal ground. The angle between the base of the ladder and the ground is 60°. The coefficient of friction between the base of the ladder and the ground is μ.

(a) Find the magnitude of the friction force that acts on the base of the ladder. (4 marks)

(b) Find an inequality that μ must satisfy. (4 marks)

7 As a particle of mass 2 kg moves, it experiences a resistance force of magnitude 6v newtons, where v m s^{-1} is the speed of the particle at time t seconds. No other forces act on the particle. At time $t = 0$ the speed of the particle is 30 m s^{-1}.

(a) Show that $\dfrac{dv}{dt} = -3v$. (2 marks)

(b) Find an expression for the velocity of the particle at time t. (4 marks)

(c) Sketch a graph to show how the velocity of the particle varies with time. (2 marks)

8 An eskimo is initially on top of his igloo and he begins to slide. Assume that the igloo is a smooth hemisphere of radius 3 m and that there is no air resistance. At time t the eskimo is moving with speed v m s^{-1} and the angle between a radius from the centre of the base of the hemisphere to the eskimo makes an angle θ with the vertical as shown in the diagram below. In the diagram the eskimo is at the point P.

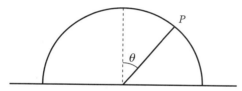

(a) Show that $v^2 = 6g(1 - \cos \theta)$. (3 marks)

(b) Find the value of θ at the point at which the eskimo leaves the surface of the igloo. (6 marks)

(c) How would your answer to (b) change if air resistance was present? (1 mark)

Answers

1 Moments and equilibrium

EXERCISE 1A

1. (a) 40 N m; (b) −300 N m; (c) 240 N m;
   (d) 21.2 N m; (e) −20.8 N m; (f) −68.9 N m;
   (g) −7.62 N m; (h) −23.1 N m; (i) −17.3 N m.
2. (a) 0.5 m; (b) 0.943 m; (c) 1.15 m.

EXERCISE 1B

1. 1.5 m from centre.
2. 270 N, 515 N.
3. (a) 196 N each; (b) 294 N; 392 N.
4. (a) 110 N; 36.8 N; (b) 7.5 kg.
5. (a) 32.7 N; 163 N; (b) 4 kg.
6. 1.05 kg.
7. Within 29 cm of the centre.
8. 120 g, 300 g.
9. (a) B; (b) 700 g.
10. (c) 374 N.
11. (b) 169 N, 490 N.
12. 196 N, 98 N.
13. (a) 2040 N, 408 N; (b) 1.56 m.

EXERCISE 1C

1. (a) 20 N m; (b) Vertical and 2 m from O.
2. 86.6 N.
3. (c) 588 N (d) 0.303.
4. (a) 35.5 N; (b) 69.5 N.
5. 56.6 N, 196 N.
6. (a) 438 N; (b) 314 N.
7. 0.520.
8. (b) 196 N.
9. (b) 0.577.
10. (c) Up to 3 m; (d) (ii) anyone can.

2 Centres of mass

EXERCISE 2A

1. 1.2 m.
2. 0.45 m.
3. 2.06 m.
4. 1.25 m, 8 kg.
5. (a) 0.1 m; (b) 0.05 m.
6. (a) (0.467, 0.167); (b) (0.13, 0.05);
   (c) (9.85, 3.68); (d) (0.2, 0.133).
7. (a) (0.7l, 0.5l); (b) (0.643l, 0.5l).
8. $m = 6$ kg, $M = 4$ kg.
9. (a) 2.5 kg; (b) 10.7 cm.
10. 4.95 cm.
11. (a) 2.2 m; (b) 216 N, 176 N.

EXERCISE 2B

1. 1.4 m.
2. 8.33 cm.
3. 26.6°.
4. 53.1°.
5. (a) 27.5 cm; (b) 17.5 cm; (c) 57.5°; (d) 40.2°.
6. 18.4°.
7. 4.45 cm, 4.73 cm, 43.3°.
8. (a) 23.2°; (b) on top edge, 7.5 cm from top left corner.
9. (a) 4 kg; (b) 10 cm.
10. (b) 4.4 cm; (c) 63.4°.
11. (a) (i) 2.6 m, (ii) 28.9°;
    (b) $x = 3$.
12. 0.84 m.
13. (b) 7.75 cm; (c) 32.3°.
14. (a) 7.75 cm; (b) 68.8°.
15. (a) (i) 28.4 cm, (ii) 41.3 cm;
    (b) $x = 28.4$ cm; (c) $x = 24.8$ cm.
16. 11.1 cm.
17. 89.0°.
18. (b) $\dfrac{113}{46}$; (c) 45.8°.

3 Energy

EXERCISE 3A

1. 4.8 J.
2. 3.75×10^{10} J.
3. 3.6 J, 14.4 J.
4. 4 725 000 J.
5. (a) 28 m s^{-1}; (b) 19.6 J.
6. (a) 7.68 m s^{-1}; (b) 1750 J.

EXERCISE 3B

1. (a) 320 000 J;
   (b) (i) 320 000 J, 25.3 m s^{-1}, (ii) 324 500 J, 25.5 m s^{-1}.
2. (a) 50 J; (b) 50 J; (c) 16.7 N.
3. (a) 62.72 J, 7.92 m s^{-1}; (b) 30.72 J, 5.54 m s^{-1}.
4. (a) 25 000 J; (b) 64 J; (c) 24 936 J; (d) 499 N.
5. (a) 1.96 J; (b) 4.43 m s^{-1}; (c) 56.25 cm.
6. (a) 1 340 000 J, 46.3 m s^{-1}; (b) 777 500 J, 1555 N.
7. (a) 300 000 J; (b) $-235\,200$ J; (c) $-64\,800$ J; (d) 20 m;
   (e) air resistance is not constant, usually some function of speed: the effect would be a greater stopping distance.
8. (a) 432 J, 108 N; (b) 1548 N.
9. (a) 249.9 N; (b) 1.45 cm.
10. 48 N, 98.0 m s^{-1}.

EXERCISE 3C

1. (a) 2450 J; (b) 9.90 m s^{-1}.
2. (a) 54 J; (b) 6.89 m.
3. (a) 416 J; (b) 440 J, 17.1 m s^{-1}; (c) 15.1 m s^{-1}.
4. (a) 630 J; (b) 750 J, 5.00 m s^{-1}; (c) 1.28 m.
5. 21.6 J; (b) 19.2 J; (c) 6.53 m.
6. (a) 32 400 J; (b) 20 640 J; (c) 12.7 m s^{-1}.
7. 286 J, 6.90 m s^{-1}.
8. (a) 37 800 J; (b) 1890 N; (c) 13.2 m.
9. 98 700 J, 13.4 m s^{-1}.
10. (a) 2695 J, 9.90 m s^{-1}; (b) 9.46 m s^{-1}.
11. (a) 48.6 J; (b) 16.5 m; (c) 42.72 J, 16.9 m s^{-1}.
12. (a) (i) 14.7 J, (ii) 10.8 m s^{-1};
    (b) 4.73 N.
13. (a) (i) 1100 J, (ii) 7 m s^{-1};
    (b) 8.93 m, 6.4 m s^{-1}; (c) 4.26 m.

14 (a) 1728 J, 6.57 m s^{-1}, no air resistance;
   (b) 2.20 m; (c) perpendicular to direction of motion.
15 (a) 431 N; (b) 25.7 m;
   (c) use a variable resistance force, model the roller coaster as a number of connected particles.

EXERCISE 3D

1 (a) 0.392 m; (b) 0.235 m; (c) 0.0392 m.
2 0.05 m.
3 156.8 N.
4 68.0 grams.
5 (a) 0.392 cm; (b) 0.196 cm; (c) 0.784 cm.

EXERCISE 3E

1 (a) 0.8 J; (b) 2.4 J.
2 7.07 m s^{-1}.
3 2.21 m s^{-1}.
5 (a) 8 J; (c) (ii) no.
6 $\lambda = \dfrac{3mg}{2}$.
7 (a) $v = \sqrt{\dfrac{800 - 70\,875x^2}{162}}$; (b) 10.6 cm.
9 (a) 12 J; (b) (ii) 2.2 J, (iii) 0.949 m.
10 (a) 0.196 cm; (b) 11.2 cm.
11 (a) 16.6 m s^{-1}; (b) (ii) 14, (iii) 29.4 m s^{-2}.
12 (a) 20 m; (b) 47.2 m; (c) (ii) 21.7 m s^{-1}.
13 (b) 7.28 m s^{-1}; (c) 14.7 m.
14 50.8 cm.
15 0.657 m s^{-1}.
16 (b) 3.32 m s^{-1}; (c) 0.963 m.
17 22.3 m s^{-1}.
18 (b) (ii) 4.51 m.

EXERCISE 3F

1 (a) 94 080 J; (b) 784 W.
2 100.8 W.
3 9×10^5 W.
4 68.1 W.
5 (a) 900 N; (b) 675 N; (c) 0.525 m s^{-2}.
6 (a) 4000 N; (b) 0.4 m s^{-2}.
7 (a) $\dfrac{75v}{4}$; (b) 11 700 W; (c) 27.1 m s^{-1}.

8 31.2 m s^{-1}, 80.2 m s^{-1}.
9 (a) $\dfrac{400v}{27}$; (b) 37.2 m s^{-1}; (c) 54.5 m s^{-1}.
10 (a) 240 000 J; (b) (ii) 40 m s^{-1}.
11 (b) (ii) 44 800 W; (c) 20.1 m s^{-1}.
12 (b) 7.5°.
13 (b) 2.1 m s^{-2}; (c) 29.6 m s^{-1}.
14 (a) 34.6 m s^{-1}; (b) 24.5 m s^{-1}.
15 (a) 200 N; (b) 427 000 J.
16 (b) 1.33 m s^{-1};
   (c) (i) motion very slow, so little air resistance,
       (ii) would accelerate for ever without air resistance.
17 (a) 30v N; (b) 19%;
   (c) small reduction leads to larger fuel savings; (d) 28.6 m s^{-1}.
18 (a) 625 N; (b) 32.7 m s^{-1}.
19 7500 W, 5 m s^{-2}.
20 (a) 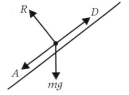 (c) 50 000 W.

There are really 4 reaction forces.

4 Kinematics and variable acceleration

EXERCISE 4A

1 (a) $83\tfrac{1}{3}$ m; (b) $v = \dfrac{t^2}{2} - \dfrac{t^3}{30}$; (c) $a = t - \dfrac{t^2}{10}$;

   (d) Increases from 0 initially, to a maximum at $t = 5$ and then decreases to 0 when $t = 10$.

2 (a) $v = 2t - \dfrac{t^2}{20}$, $a = 2 - \dfrac{t}{10}$; (b) $t = 20$;
   (c) 20 m s^{-1}; (d) 267 m.

3 (b)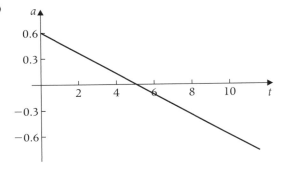

4 (a) $0 \leq t \leq 15$; (b) 112.5 m;
(c)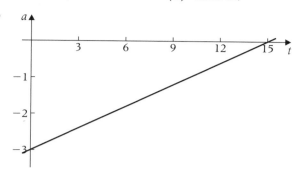

5 243 m.

6 (a) $v = 4 - 2e^{-t}$; (b) $a = 2e^{-t}$.
(c) decreases from 2, tending to zero as $t \to \infty$.

7 (a) $v = 0.2 \cos(0.5t)$; (b) 0.2 m s^{-1};
(c) -0.0841 m s^{-2}; (d) $-0.1 \leq a \leq 0.1 \text{ m s}^{-2}$.

8 (a) 0 m; (b) $0 \leq t \leq 800\pi$;
(c) 300 m; (d) $0.000\,234 \text{ m s}^{-2}$.

9 (a) $A = 15$; (b) $k = \dfrac{4}{3}$;
(c)

10 (a) 4 m; (b) 6 m s^{-1}.

11 (a) $0 \text{ m s}^{-1}, 40 \text{ m s}^{-1}$,
(b)

12 (b) $-(kU+g)$ m s^{-2}

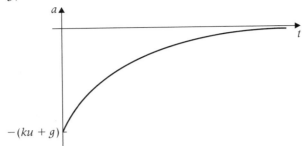

13 (a) $v = 5t - \dfrac{t^3}{5}$, $a = 5 - \dfrac{3t^2}{5}$;   **(b)** $0 \leq t \leq 5$ s, 31.25 m;

(c) Decreases from 5 m s^{-1} to -10 m s^{-2}, 9.62 m s^{-1}.

EXERCISE 4B

1 (a) $v = \dfrac{t^3}{300}$;   **(b)** $\dfrac{5}{12}$ m s^{-1};   **(c)** $\dfrac{25}{48}$ m.

2 12.8 m s^{-1}, 55.5 m.

3 38.3 m.

4 11.1 m s^{-1}, 42.1 m.

5 (b) 17.5 m s^{-1};   **(c)** 183 m.

6 (a) $v = 3t - \dfrac{t^2}{20}$;   **(b)** $t = 30$ s, $v = 45$ m s^{-1};

(c) 900 m.

7 (b) $s = t^2 - 2e^{-t} + 2t$.

8 (a) $a = 1.8 \cos(3t)$ m s^{-2};   **(b)** $x = 1 - 0.2 \cos(3t)$ m;

(c)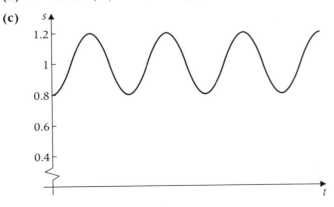

9 (a) $v = 4.9(1 - e^{-2t})$, $s = 2.45e^{-2t} + 4.9t - 2.45$;

(b) [graph of v vs t, approaching 4.9] (c) 95.6 m.

10 18.7 m.

11 (a) $v = \dfrac{2}{5\pi} \sin(100\pi t)$ m s^{-1}; (b) $\dfrac{2}{5\pi} = 0.127$ m s^{-1};

(c) 0.000 811 m.

12 (a) $x = \dfrac{3}{2} - \dfrac{3}{2}\cos(2t) - \sin(2t)$; (b) $\sqrt{13}$ m s^{-1}.

EXERCISE 4C

1 (a) $\mathbf{v} = 2t\mathbf{i} + (12t - 1)\mathbf{j}$, $\mathbf{a} = 2\mathbf{i} + 12\mathbf{j}$;

(b) $\mathbf{r} = 11\mathbf{i} + 96\mathbf{j}$, $\mathbf{v} = 8\mathbf{i} + 47\mathbf{j}$.

2 (a) $\mathbf{v}_A = 30\mathbf{i} + (6t - 120)\mathbf{j}$, $\mathbf{v}_B = 20\mathbf{i} + 40\mathbf{j}$;

(b) $v_B = 44.7$ m s^{-1};

(c) $t = 30$ s, 756 m.

3 $\mathbf{v} = 2t\mathbf{i} + 2\mathbf{j}$, $\mathbf{a} = 2\mathbf{i}$.

4 (a) $\mathbf{v} = \left(4 - \dfrac{2t}{5}\right)\mathbf{i} + 10\mathbf{j}$; (b) $t = 20$ s, 200 m;

(c) $t = 10$ s, 10 m s^{-1}; (d) $\dfrac{2}{5}$ m s^{-1} due west.

5 (a) $\mathbf{v} = (2t - 8)\mathbf{i} + (6t^2 - 10t + 6)\mathbf{j}$;

(b) $\mathbf{F} = 6\mathbf{i} + (36t - 30)\mathbf{j}$;

(c) $t = 4$ s, $-14\mathbf{i} + 72\mathbf{j}$.

6 (a) $t = 2$ s, $v = 20.2$ m s^{-1}; (b) $t = 5$ s, $\mathbf{a} = -\mathbf{j}$.

7 $38.3\mathbf{i} + 16.7\mathbf{j}$.

8 (a) 3.2 kg; (b) $28.3\mathbf{i} + 46.9\mathbf{j}$.

9 $\mathbf{v} = (2 + t)\mathbf{i} + (5 + 0.1t^2)\mathbf{j}$, $\mathbf{r} = \left(2t + \dfrac{t^2}{2}\right)\mathbf{i} + \left(5t + \dfrac{t^3}{30}\right)\mathbf{j}$.

10 (a) $\mathbf{v} = (3 + t^2)\mathbf{i} + \left(6 - \dfrac{5t^2}{2}\right)\mathbf{j}$; (b) $\mathbf{r} = \left(3t + \dfrac{t^3}{3}\right)\mathbf{i} + \left(6t - \dfrac{5t^3}{6}\right)\mathbf{j}$.

11 (a) $\mathbf{v} = 4t\mathbf{i} - \dfrac{t^2}{2}\mathbf{j}$; (b) $\mathbf{r} = (5 + 2t^2)\mathbf{i} - \left(10 + \dfrac{t^3}{6}\right)\mathbf{j}$.

12 (a) $\mathbf{v} = \begin{bmatrix} 4t \\ 5 \end{bmatrix}$; (b) 0.5 kg;

(c) (i) $\mathbf{a} = \begin{bmatrix} 4 \\ 2t \end{bmatrix}$, (ii) $\mathbf{v} = \begin{bmatrix} 40 \\ 80 \end{bmatrix}$.

13 (a) $\mathbf{v} = (2t - 6)\mathbf{i} + t^2\mathbf{j}$; (b) $t = 3$ s; (c) $\mathbf{a} = 2\mathbf{i} + 2t\mathbf{j}$, not constant as it depends on t.

14 (a) $\mathbf{a} = \frac{1}{2}t\mathbf{i} - \frac{5}{2}\mathbf{j}$; (c) $\mathbf{r} = \frac{t^3}{12}\mathbf{i} + \left(6t - \frac{5t^2}{4}\right)\mathbf{j}$.

15 (a) $\mathbf{v} = (3t^2 - 6t)\mathbf{i} + (4 + 4t)\mathbf{j}$;
   (b) $a = \sqrt{(6t - 6)^2 + 4^2}$; (c) $t = 1$ s, $a = 4$ m s^{-2}.

16 (a) $\mathbf{v} = -4 \sin t\mathbf{i} + \left(\frac{7}{2} - 3 \cos t\right)\mathbf{j} + \frac{t}{2}\mathbf{k}$;
   (b) $\mathbf{r} = 4 \cos t\mathbf{i} + \left(\frac{7t}{2} - 3 \sin t\right)\mathbf{j} + \frac{t^2}{4}\mathbf{k}$;
   (c) 2.57 m.

17 (a) $\mathbf{v} = -2 \sin\left(\frac{t}{20}\right)\mathbf{i} + 4 \cos\left(\frac{t}{20}\right)\mathbf{j}$;
   (b) north; (c) 20π s.

18 32 m s^{-1}, 256 m s^{-2}.

19 (a) $\mathbf{r} = -4\mathbf{i}$, $\mathbf{v} = -6\mathbf{j}$, $\mathbf{a} = 16\mathbf{i}$; (b) 12 m s^{-2}, 16 m s^{-2}.

20 (a) Tends towards $-40\mathbf{j}$;
   (b) $\mathbf{r} = 700(1 - e^{-0.1t})\mathbf{i} + (400 - 400e^{-0.1t} - 40t)\mathbf{j}$.

21 $t = 0$ s and $t = \pi$ s.

5 Circular motion

EXERCISE 5A

1 (a) $\frac{1}{60}$ rpm; (b) $\frac{\pi}{1800}$ rad s^{-1}.

2 $\frac{\pi}{3}$ rad s^{-1}.

3 3640 m s^{-1}.

4 29 900 m s^{-1}.

5 52.4 m s^{-1}, 13 700 m s^{-2}.

6 35.5 N.

7 2.5 m.

8 1.54 m s^{-2}.

9 20.9 m s^{-1}, 2190 m s^{-2}.

10 (a) $\frac{\pi}{3}$ rad s^{-1}; (b) $\frac{2\pi}{3}$ m s^{-1}, $\frac{\pi}{2}$ m s^{-1}.

11 (b) 1020 m s^{-1}.

12 (b) 0.90 m; (c) 0.628 m s^{-2}.

EXERCISE 5B

1. 1750 N.
2. 3.6 N towards the centre of the circle.
3. 72 N.
4. 11.6 m s^{-1}.
5. 576 N.
6. $2\pi\sqrt{\dfrac{a^3}{k}}$.
7. 38.1 N.
8. 3290 N, 0.336.
9. 6.26 rad s^{-1}.
10. (a) Coin slides; (b) 4.4 cm; (c) No change.
11. 0.227, 1.19m N.
12. (b) 6 m s^{-1}; (c) 3 rad s^{-1}; (d) 72 N.
13. 0.340.
14. 18.8 m s^{-1}.
15. 1.98 rad s^{-1}.
16. (a) 49.3 m s^{-2}, 25.2 N; (b) 0.639 s.
17. (a) 0.0392 N; (b) 8.85 rad s^{-1};
    (c) reduced to $\frac{1}{4}$ of previous value.
18. (a) 125 m s^{-2}; (c) path is a tangent to the end of the strip.
19. 0.33.
20. (a) 6000 N; (b) $\mu \geqslant 0.510$; (c) 21.7 m s^{-1}; (d) air resistance.
21. (a) 98 N; (b) 3.70 m s^{-1};
    (c) moves at a tangent to the circle.
22. (a)

    (b) 16.3 m; (c)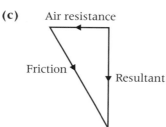

EXERCISE 5C

1. (a) 20.9 N, 4.57 rad s^{-1};   (b) 50 N, 66.9°;   (c) 4.9 m, 39.2 N.
2. 20.0 N, 4.47 rad s^{-1}.
3. 82.9 N, 65.5 cm.
4. (a) 48 N;   (b) 35.2°;   (c) 1.73 m.
5. 56.6 N, 1.68 m s^{-1}.
6. (b) 2.56 rad s^{-1}.
7. 0.959 rad s^{-1}.
8. (a) (i) 33.9 N;   (b) 77°.
9. (a) 160 m s^{-2};   (b) 22.6 N;   (c) 309 N.
10. (c) 18.4 N.
11. $\dfrac{g}{l\omega^2}$.
12. $\dfrac{g}{2\pi^2}$.
13. $v = \dfrac{\sqrt{15ag}}{2}$, $R = 4\,mg$.
15. (a) 10 000 N;   (b) $\alpha = \tan^{-1}\left(\dfrac{v^2}{rg}\right)$, $L = m\sqrt{\dfrac{v^4}{r^2} + g^2}$;
    (c) reduce r then α increases and L increases.
16. (a) (i) 453 N,   (ii) 5.66 m s^{-2},   (iii) 3.36 m s^{-1};
    (b) 0.553 s, due to the actual size of the child;
    (c) 2.73 m, between 2 and 2.73 m from O.
17. (b) $\dfrac{gr^2}{\sqrt{l^2 - r^2}}$;   (c) $\dfrac{ml}{2}\left(\dfrac{g}{\sqrt{l^2 - r^2}} + \dfrac{V^2}{r^2}\right)$.
18. (a) 19.7 m s^{-2};   (b) $R = 1380$ N, $F = 686$ N.
19. (a) $\tan \alpha = 5$;   (c) $3.83 \leqslant v \leqslant 4.43$ m s^{-1};
    (d) 0.0735 J;   (e) rotation of the coin.

6 Circular motion with variable speed

EXERCISE 6A

1. 1.58 m s^{-2}.
2. (a) 2.46 m s^{-2};   (b) 1.86 m s^{-2}.
3. 2.27 m s^{-2}.
4. 19.0 m s^{-1}.
5. 110 m.
6. (a) $\dfrac{25a}{2}$;   (b) $\dfrac{V^2\sqrt{626}}{a}$.
7. 8.74 m s^{-2}.

EXERCISE 6B

1. (a) 3.97 m s^{-1}, 932 N; (b) 799 N.
2. (a) 4.65 m s^{-1}, 0.05 N; (b) 2.99 N.
3. (a) 556 N; (b) $48.0°$.
4. 2.56 m s^{-1}, 435 N.
5. 3230 N, 1670 N.
6. 2.21 m s^{-1}.
7. 4.58 rad s^{-1}.
8. $3mg \sin \theta$.
9. $\dfrac{a}{3}$.
10. $R = m\left(\dfrac{u^2}{a} + g(3\cos\theta - 2)\right)$
11. (a) $\sqrt{\dfrac{4rg}{3}}$; (b) (i) $\dfrac{mg}{3}$, (ii) $\dfrac{29mg}{6}$.
12. (a) $\sqrt{\dfrac{4gr}{15}}$; (b) $\dfrac{79mg}{15}$.
14. (b) $\theta = \cos^{-1}\left(\dfrac{u^2 + 2gl}{3gl}\right)$.
15. (b) (i) $\dfrac{11mg}{15}$, (ii) $\dfrac{23mg}{30}$.
16. (b) (ii) $34°$, (iii) 530 N, (iv) no air resistance.
17. (a) $2gr$; (b) (iii) $mg(2 + 3\sin\theta)$; (d) $\dfrac{g}{2}$.
18. (b) $5gr$; (c) $k = 2.5$; (d) consider fraction/air resistance.
19. (a) $\sqrt{\dfrac{rg}{2}}$.
20. (a) $\sqrt{5gl}$; (b) $(\sqrt{5} - 2)\sqrt{gl}$

7 Application of differential equations in mechanics

EXERCISE 7A

1. (a) $v = 20e^{-2.5t}$; (b) $s = 8(1 - e^{-2.5t})$; (c) 8 m.
2. (a) $v = \dfrac{10}{5t + 2}$; (b) $v = \dfrac{20}{10t + 1}$.
3. (a) $v = 80e^{-\frac{t}{2}}$; (b) $s = 160(1 - e^{-\frac{t}{2}})$; (c) 120 m.
4. (a) $v = 20e^{-2.5t}$; (b) 7.95 m.
5. 167 s.
6. 61.2 m s^{-1}.

152 Answers

7. (a) 0.467 s; (b) 3.18 m s^{-1}.
8. (a) $v = \dfrac{600}{t+20}$; (b) 100 s.
9. (a) $20e^{-\frac{t}{5}}$ m s^{-1}; (b) $100(1-e^{-\frac{t}{5}})$ m.
11. (b) v tends to zero.
12. (a) $v = Ue^{-4t}$; (b) $s = \dfrac{U}{4}(1-e^{-4t})$.
13. (a) $v = 10e^{-\frac{t}{8}}$, $s = 80(1-e^{-\frac{t}{8}})$.

Exam style practice papers

MM2A

1. (c) 130.7 N.
2. (a) (i) $\mathbf{v} = 2\cos t\mathbf{i} - 3\sin t\mathbf{j}$, (ii) 2.85 m s^{-1};
   (b) (ii) 2 m s^{-2}.
3. (a) 400 J;
   (b) (i) 392 J, (ii) 1.26 m s^{-1}, (iii) 4.08 m.
4. (b) 1.30 m s^{-1}.
5. (b) 3.55 cm; (c) 33.6°.
6. (b) $v = \dfrac{14}{7t+2}$; (c) $t = \dfrac{5}{7}$ s.
7. (b) 132°.

MM2B

1. (a) 7.2 J; (b) 11.6 m s^{-1};
   (c) 6.84 m; (d) no air resistance.
2. (a) (i) $\mathbf{v} = 21t^2\mathbf{i} + (16t-5)\mathbf{j}$, (ii) 0.3125 s;
   (b) (i) $\mathbf{a} = 42t\mathbf{i} + 16\mathbf{j}$, (ii) $t = 0$.
3. (a) 8232 N; (b) 18.5 m s^{-1}.
4. (b) 2.736 m; (c) 15.7°;
   (d) yes, because the position of the centre of mass would change.
5. (a) 96 000 W; (b) 1 m s^{-2}; (c) 9.79 m s^{-1}.
6. (a) 33.9 N; (b) $\mu \leq 0.289$.
7. (b) $v = 30e^{-3t}$;
   (c) graph showing exponential decay from 30 to zero.
8. (b) 48.2°; (c) the angle would increase.

Index

acceleration 68–71, 75–79, 89
   and circular motion 92–93, 111
     at variable speed 112–113, 125
   and differential equations 126
   in two or three dimensions 81–84
acceleration-time graphs 71
angles, forces at 38–42
angular speed 91–93, 111
   and horizontal motion 95–96
anticlockwise turning effect 2, 15
area, centre of mass from 22–23

balance 5, 16
bodies
   composite, centre of mass 22–25, 30
   uniform 4–6, 22

centre of mass 16–20
   composite bodies 22–25, 30
   suspended laminas 24–25
circular motion 91–93
   conical pendulum 101–104
   horizontal 95–97, 111
     with vertical motion 101–104
   at variable speed 112–114
   vertical 115–118
     with horizontal 101–104
clockwise turning effect 2, 15
components of forces 38–42
   in circular motion 113–114
composite bodies, centre of mass of 22–25, 30
conical pendulum 101–104
conservation of energy 115–118
constant acceleration 68
coordinates of centre of mass 18–19

differential equations 126–134
differentiation 69–71, 73, 81–82, 89
displacement 69–71, 75–79, 89
   see also distance; position vectors
displacement-time graphs 69
distance
   and moments 1–2, 10, 15
   and turning effect 1–2
   and work 33–36, 57
     and forces at angles 38–42
     and variable forces 49–52
   see also displacement

elastic potential energy 49–52, 65
elastic strings 47, 51–52
energy 32–33, 65
   potential energy 40–42, 65
     elastic 49–52, 65
     in vertical circular motion 115–118
   in vertical circular motion 115–118
   and work 33–36, 65
     forces at angles 38–42
     and power 57
     variable forces 49–52
EPE (elastic potential energy) 49–52, 65
equilibrium 4–6, 9–11
   and stretched bodies 47–48, 52

forces
   in circular motion 92, 113–114
     conical pendulum 101–104
     horizontal 95–97
     vertical 115–118
   and Hooke's law 47–48, 65
   and moments 1–3, 15
     and equilibrium 4–6, 9–11, 15
   reaction forces 4–6, 10, 96
   resultant force 4, 15
     in circular motion 95, 111, 113–114
   and rigid body model 1
   and work 33–36, 49
     forces at angles 38–42
     and power 57–60
     variable forces 49–52, 65
   see also friction; tension
friction
   in circular motion 96–97, 113–114
   and moments 10–11
   and work 35–36, 39–40
   see also resistance forces

graphs
   of motion 69, 71, 78
   of work done 49
gravitational potential energy 40–42, 65
   in stretching problems 51–52
   see also potential energy
gravity
   and kinetic energy 33
   and vertical circular motion 115
   and work 34, 36, 40

Hooke's law 47–48, 65
horizontal circular motion 95–97, 111
   with vertical motion 101–104

integration 73, 75–79, 83–84, 89
   and differential equations 126–130

key point summaries
   centre of mass 30
   circular motion 111
      variable speed 125
   differential equations 134
   energy 65
   kinematics 89
   moments and equilibrium 15
kinematics 68–71, 75–79, 89
   in two or three dimensions 81–84
  see also motion
kinetic energy 32–33, 65
   in vertical circular motion 115–116, 118
   and work 33–36, 65
      forces at angles 39–42
      and power 57

laminas 22, 23–25
line of action 2–3

mass
   and kinetic energy 32
   see also centre of mass
modulus of elasticity 47
moments 1–3, 15
   and centre of mass 16–20, 22–25, 30
   and equilibrium 4–6, 9–11
motion 68–71, 75–79, 89
   circular *see* circular motion
   and differential equations 126–134
   and energy 32–36
      forces at angles 38–42
      and power 57–60
   in two or three dimensions 81–84

particles 1, 4
   systems of, centre of mass 16–20
position vectors 81–84, 89, 93
potential energy 40–42, 65
   elastic 49–52, 65
   in vertical circular motion 115–118
power 57–60, 65

radial components 112–114, 125
reaction forces 4–6, 10
   in horizontal circular motion 96
resistance forces 35–36, 40, 42
   and differential equations 127–130
   and power 58–60
   see also friction
resultant force 4, 15
   in circular motion 113–114
      vertical 95, 111
rigid bodies 1

separation of variables 126–130
speed
   angular speed 91–93, 111
   in circular motion 93, 95–96, 111
      variable 112–114
   and kinetic energy 32, 34–35
      forces at angles 39–42
   and power 57–60
   see also velocity
springs 47–48, 49–51, 65
stretching 47–48, 49–51, 65
strings, elastic 47, 51–52
suspended laminas, centre of mass of 24–25

tangential components 112–114, 125
tension
   and circular motion 96, 115–116
      conical pendulum 101–103
   and Hooke's law 47–48, 65
   and moments 9–11
time (in motion problems) 68–90
   and differential equations 126–134
turning effect 1–2, 15

uniform bodies 4–6, 22

variable acceleration 68–71, 75–79, 89
   in two or three dimensions 81–84
variable forces 49–52, 65
variable speed 112–114
variables, separation of 126–130
vectors, position 81–84, 89, 93
velocity 69–71, 75–79, 89
   and circular motion 92–93, 111
   and differential equations 126–130, 134
   in two or three dimensions 81–84
   see also speed
velocity–time graphs 69, 71, 78
vertical circular motion 115–118, 125
   with horizontal 101–104

work 33–36
   forces at angles 38–40, 42
   and power 57–60
   variable forces 49–52